WORKBENCH GUIDE TO
ELECTRONIC PROJECTS
YOU CAN BUILD
IN YOUR SPARE TIME

Other Books by Carl G. Grolle

Complete Guide to Electrical and Electronic Repairs

Electronic Technician's Handbook of Time-Savers and Shortcuts

Electronic Workshop Manual and Guide

Grolle's Complete Guide to Electronic Troubleshooting

WORKBENCH GUIDE TO ELECTRONIC PROJECTS YOU CAN BUILD IN YOUR SPARE TIME

Carl G. Grolle
and
Michael B. Girosky

Parker Publishing Company, Inc.
West Nyack, New York

© 1981 by

PARKER PUBLISHING COMPANY, INC.

West Nyack, New York

Library of Congress Cataloging in Publication Data

Grolle, Carl G.
 Workbench guide to electronic projects you can build
in your spare time.

 Includes index.
 1. Electronics—Amateurs' manuals. I. Girosky,
Michael B. II. Title
 TK9965.G77 621.381 81-2169
 ISBN 0-13-965269-8 AACR2

Printed in the United States of America

This Book Is Dedicated to

A loving family...Irene, Eleonore, Lillian and George

C.G. Grolle

My loving wife Marj for her help,
encouragement and understanding

M.B. Girosky

The Many Ways
This Projects Book
Will Be Helpful to You

This book is written for the person who enjoys building electronic projects from scratch but doesn't want to spend a lot of time and money chasing after a long list of hard-to-find parts. Most of the projects featured in this book are inexpensive and require only a minimum number of standard components. How many times have electronics enthusiasts started on a complex electronic project with great intentions only to get bogged down in the complexities of mind-boggling circuits and marginal design? These fiascos are not only expensive, but downright frustrating.

You'll like these projects — they have been selected as practical, worthwhile, interesting and fun to build. Many have detailed PC board patterns that are ready to be copied. Or, if you prefer, it's even easier to build these circuits on perfboard.

A new idea in project building called the *Build-O-Matic Guide* is featured in this book. Every project has special Build-O-Matic Guide identification points to make successful construction as simple as ABC. Just match up the Build-O-Matic Guide letters and you can't go wrong. It's a concept so simple and so worthwhile that you'll wonder why all electronic projects are not identified in this manner.

This book is divided into the following sections:

1. Getting the Most Value from Electronic Projects
2. Projects that Will Make Money for You
3. Energy-Saving Projects You Can Build

4. Testers for Checking Solid State Components
5. Projects to Make Life Easier
6. Money-Saving Electronic Projects
7. Projects Just for Fun

Each section is jam-packed with interesting electronic projects that you will want to make. Complete instructions are given for producing professional printed circuits quickly ... and easily. In addition you will find straightforward plans for fast chassis construction using ony a few common sheet metal tools.

Section 4, *Testers for Checking Solid State Components*, describes a wide variety of electronic test instruments that are guaranteed to find the worst electronic bugs. You'll find testers for checking diodes, zener diodes, rectifiers, bipolar transistors, unijunction transistors, silicon controlled rectifiers, triacs, quadracs, and many more. With this assortment of testers on hand you can buy surplus solid state components at rock bottom prices, easily identifying the good from the bad.

If you are interested in projects for fun, you'll like the variable intensity and duration strobe light project. Build this little beauty for about one-third the price of a ready-built model. How about an emergency siren that will really attract attention? Maybe you are looking for a dazzling color organ for brilliant light displays. You'll find plans and projects in this book that will make you the envy of the neighborhood.

Energy saving is on everybody's mind these days and rightfully so. There are numerous projects that will literally pay for themselves in a month or two after completion. For example, the hot water tank is a common item found in everyone's home; yet it's one of the largest energy wasters around. This book will show you in detail how to reduce your hot water energy consumption using a simple electrical project that can save a hundred or more dollars each year. The fireplace is another tremendous energy waster. It's a crime to let all that heat energy go up the chimney. Why not build an electrical system that will capture some of this precious heat—even during a complete power failure.

Let Section 5, *Projects to Make Life Easier,* work for you. It's human nature to try to make your life a little easier and simpler. Let electronics be your servant. Wouldn't it be great to jump into bed ready for a good night's sleep and have your stereo turn itself off automatically after you have fallen asleep? Or how about a lamp that will turn on and off like magic with just a gentle touch of your hand? This book shows you how to let electronics give your life a little luxury.

We could go on and on about some of the other extraordinary features in this book, but enough has been said. We believe you will like it, have a lot of fun with it, *and* save yourself both time and money. In short, your investment in this book will pay off in *many* ways.

C.G. Grolle

M.B. Girosky

Contents

1

Getting the Most
Value from
Electronic Projects

1—1: USING THE BUILD-O-MATIC GUIDE

What It Does

White key letters superimposed on black triangles identify the Build-O-Matic Guide symbols on each project schematic. These Guide letters are used to simplify project construction. For example, suppose there are a number of schematic lines identified with Build-O-Matic Guide letter "A." All the builder has to do is connect all the "A" lines together at a single terminal strip or part lug. All letter "B" points would be wired together . . . then "C's," etc. Wiring each project is just as easy as matching letters. All lines with no Build-O-Matic Guide letters simply connect between the two parts on the schematic.

Input and output connections to printed circuits are also identified with Build-O-Matic Guide letters. Once the PC board is constructed, just hook up all the like Build-O-Matic letters together to make a goof-proof project.

Here's the Schematic

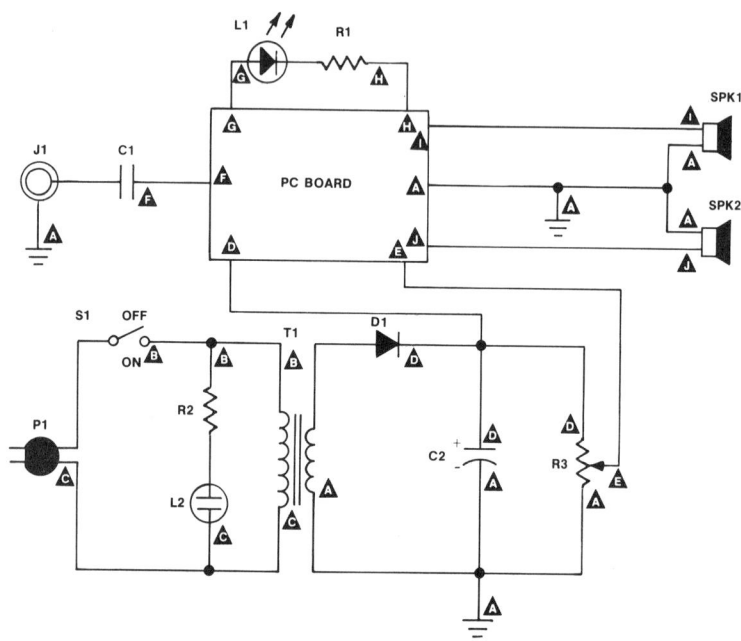

**FIGURE 1—1:   BUILD-O-MATIC GUIDE SYMBOLS ON A
SAMPLE SCHEMATIC**

Helpful Construction Hints

Suppose we are wiring the project shown in Figure 1—1. We'll assume that the printed circuit has been completed and all chassis parts have been mounted. Let's connect C1 to the circuit. Since the input side of C1 has no Build-O-Matic Guide letter it is wired directly to the points shown on the schematic. For example, the left side of C1 is soldered to the center terminal of J1. The right side of C1, marked F, is connected to the PC board input connection lettered F. If the physical distance between J1 and PC terminal F is too long for the capacitor leads, simply mount C1 on a terminal strip using hookup wire to complete the connection.

Now let's tackle another part of the schematic identified with Build-O-Matic Guide letters. Notice that the outside connection of J1 is lettered "A." There are a number of other points on the schematic that are also identified with the letter "A." To be specific, these are the bottom of T1's secondary winding, the bottom of C2, the bottom of R3, the bottom of SPK1 and the top of SPK2, foil pad A of the PC board and chassis ground. All of these points are wired together. They could be connected together with hookup wire or they could each be fastened to a chassis ground point making the chassis act as the conductor.

Let's do one more example using Build-O-Matic Guide identification. Find Build-O-Matic Guide letter "B." The right side of S1, the top of T1's primary winding, and the top of R2 are all identified with letter "B." A typical way to wire these parts would be as follows: since S1's right terminal lug is physically solid it would be a good tie point for connecting the R2's lead and the wire from T1's primary winding. Once this connection is soldered, all Build-O-Matic Guide letter B connections are complete. Of course, if the part leads are not long enough to reach the S1 lug just use a terminal strip conveniently located to make the connections.

1—2: PRINTED CIRCUIT CONSTRUCTION MADE SIMPLE

What it Does

This section describes in detail how a printed circuit can be quickly and easily made from an exact scale layout. Building many of the projects in this book is simplified by the use of printed circuit (PC) construction. You will find an exact scale foil pattern included for every project that would benefit from PC construction.

Although there are many ways to make a printed circuit, the technique explained in this section is probably the least expensive and most straightforward for one-of-a-kind type projects. You will find that the number of PC materials needed are minimal and can be used over and over for any of the projects.

Materials to Use

Artist's paint brush	Paint thinner
Asphaltum resist	Pencil carbon paper
Copper clad board	Pencil #2
Drill bit, 1/32″	Resist pen
PC etchant	Steel wool, fine

The PC Layout You'll Need

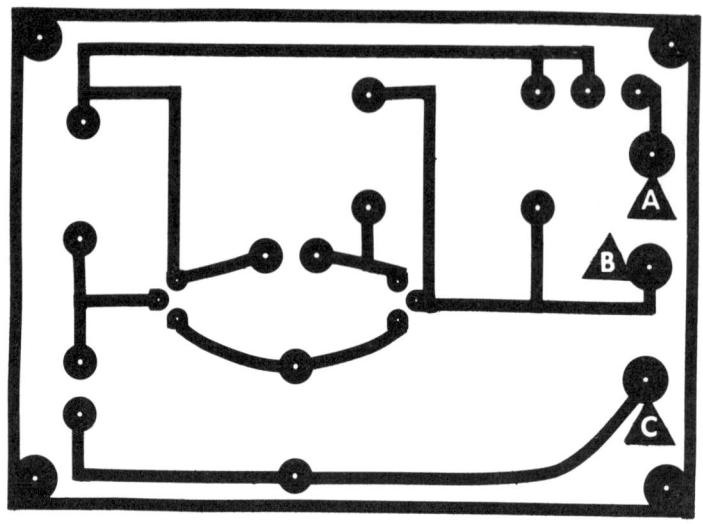

FIGURE 1—2A: TYPICAL PC LAYOUT (ACTUAL SIZE)

Helpful Construction Hints

The PC layout shown in Figure 1—2A is not meant to be a project in this book but only an example of a typical PC layout. The first step in PC construction is to transfer the PC layout to the copper-clad board. Cut a PC board to the size of the layout pattern. Next clean the copper-clad foil with steel wool until it is bright shiny copper. Place the PC board foil side up directly underneath the PC layout. Insert pencil carbon paper between the page and the PC board with the carbon side toward the copper foil. Carefully trace the PC layout with a #2 pencil, being sure to include all details. Do not let the page, carbon paper, or PC board shift while making the transfer.

Remove the PC board from the book and inspect it for any missing items. You can draw in missing lines with a pencil. The next step is to brush asphaltum resist paint on the PC board within the areas outlined by the carbon paper. A small brush and a little patience will do wonders. For very small areas, a resist pen, available at many electronic supply stores, will do a good job. After the resist has dried, you are ready for the etchant phase of the PC construction.

Pour the etchant solution in a shallow glass or plastic container. Place the PC board foil side down in the etchant making sure it is entirely covered. Agitate the PC board from time to time to speed up the etching process. The copper should be completely removed within a half hour or so. Rinse the PC board with water and inspect it to make sure that all the copper has been etched away. Save the etchant solution. It can be used a number of times before it becomes ineffective.

Now remove the resist with paint thinner. If everything has gone right your PC board should be a work of art. Again, inspect it carefully, looking for copper bridges between foil areas. These bridges can usually be removed by scraping with a sharp knife. Lightly center-punch all holes to be drilled. Drill each hole with a 1/32″ drill bit at high speed. After all holes have been drilled, clean off the burrs and shine up the copper with fine steel wool. Wipe away any steel wool residue. You are now ready to insert parts into the PC board.

Although this discussion on PC construction appears to be quite lengthy, the actual work is fairly simple, straightforward and kind of fun. It doesn't take long to make a PC board once you get the hang of it. After you have made several boards it's easy to make any of the layouts in this book in less than an hour.

In summary, here's a list of essential steps in PC board construction:

1. Cut and steel wool PC board
2. Transfer PC pattern from book to copper-clad board using carbon paper
3. Paint and/or pen resist on PC board
4. Etch board
5. Remove resist
6. Center-punch and drill PC board
7. Steel wool and clean PC board
8. Mount parts

For other techniques in printed circuit construction refer to the author's book, *Electronic Technician's Handbook of Time-Savers and Shortcuts*, available from Parker Publishing Co. Inc., West Nyack, N.Y. 10994.

How It Works

The PC resist coating gives a protective coat to the copper-clad. When the board is submerged into the etchant solution, a chemical reaction takes place causing the exposed copper foil to be slowly eaten away. Eventually, all the copper will be dissolved except for those areas protected by the resist coating.

After the PC board is etched and removed from the solution, the resist is dissolved by paint thinner, exposing the remains of the original copper foil. The copper-clad that is left forms the conductor paths and soldering islands for the electronic parts. It works like a charm.

1—3: IT'S EASY TO BUILD A PROJECT CHASSIS

Many electronics enthusiasts house their projects in a ready-made chassis or enclosure. This can be an expensive proposition, particularly for the prolific project builder. Not only are commercial enclosures costly, but they are often not suited to the needs of the hobbyist. A less costly approach is to design and build your own chassis. Using simple hand tools, you can construct a professional looking enclosure with minimal time and energy. Following the two basic steps, design and fabrication, will guarantee your success in making a project chassis.

DESIGN: A simple minibox is the ideal choice for the beginning chassis builder. (See Figure 1—3A.) Avoid complicated and intricate designs that are difficult to fabricate until you become more proficient. You must first consider the size of the enclosure. The chassis should be compact, yet provide ample space for wiring and servicing.

A basic enclosure consists of two parts—a main body and a removable dust cover. All component parts should be housed in the main body. This eliminates the need for interconnecting wires between the chassis and the dust cover. After several sketches are made, and a final size is decided upon, construct a working model out of paper. Heavy construction paper or lightweight cardboard may be used. Cut, bend, and fabricate the pattern as you would the finished product. Any design flaws, such as impossible bends, will

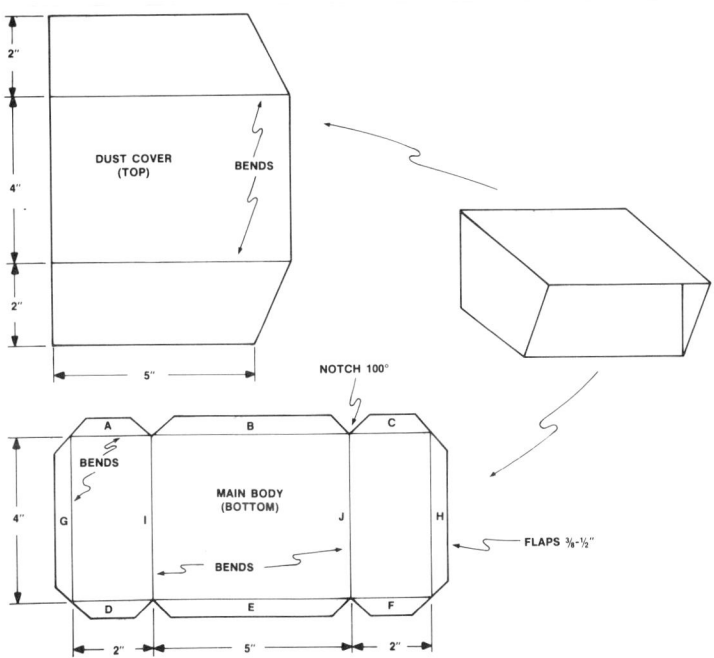

FIGURE 1—3A: MINIBOX BLUEPRINT

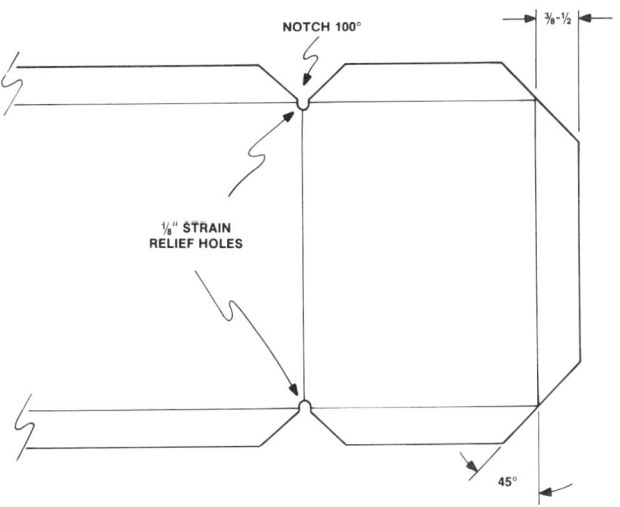

**FIGURE 1—3B: MAIN BODY CHASSIS PATTERN
(PARTIAL VIEW)**

be readily apparent. Once you are satisfied with your design, carefully transfer the pattern to the sheet metal with a scribe or well sharpened pencil.

FABRICATION: Sheet steel or aluminum of 22-24 gauge thickness can be used. Although it is more costly, aluminum is recommended because of its ease of fabrication with hand tools.

Center-punch and drill all holes in the sheet metal. For larger holes, a progression of small to larger drills is recommended. This technique prevents the holes from becoming elongated or egg shaped. Holes larger than ½″ are best made with a chassis punch, fly cutter, or hole saw; ⅛″ holes drilled at each corner will provide strain relief and prevent buckling during the bending operation. (See Figure 1—3B.)

After all holes are made, notches are cut into the main body foldlines. (See Figures 1—3A and 1—3B.) These notches should be cut at approximately 100-degree angles. This allows for sufficient clearance, and compensates for any slight irregularities incurred in the forming process. These cuts can be made with aviation snips or a fine-tooth hacksaw blade. Be sure to remove any burrs resulting from drilling and cutting. Then, fine finish all edges and openings with a medium grade abrasive paper.

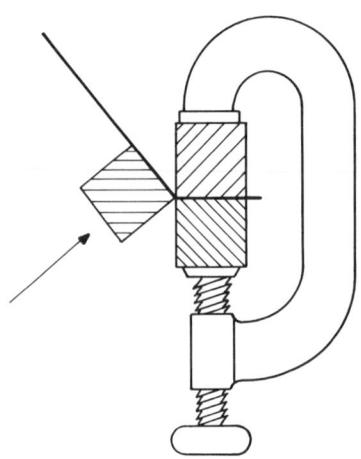

FIGURE 1—3C: BENDING SHEET METAL USING WOOD STRIPS

Since most home hobbyists do not have access to sheet metal forming equipment, here is an easy and inexpensive alternative. Using C clamps and 1" by 1" strips of wood, you can bend the metal to the desired angles. Cut the 1" by 1" wood strips the exact length of the bend. Clamp the strips securely above and below the bend, as illustrated in Figure 1—3C. Using another strip of wood, bend the metal to the desired angle.

For ease of fabrication, the following order of bending is suggested. (Refer to Figure 1—3A.) Begin by bending flaps A, B, and C simultaneously; follow with D, E, and F, and finish with flaps G and H. Complete the bending process for the main body by making 90-degree angles at I and J.

Form the dust cover using the same outlined techniques. Install and fasten the dust cover with several small self-tapping screws. With a little practice, you will soon become an expert in chassis construction.

2

Projects That Will Make Money for You

2—1: AUDIO TEST JIG

What It Does

The audio test jig is a compact servicing unit that provides quick access to a stereo speaker system and two one-watt audio amplifiers. The bench technician can conveniently work on audio problems in car radios, stereos, tape players, televisions and other kinds of equipment that are often removed for servicing minus their speakers.

A flick of a switch provides instant selection of speaker impedances. A few test lead connections are all that are necessary for substitute audio amps. You'll probably find this neat little jig to be one of your most popular servicing aids.

Here's the Schematic

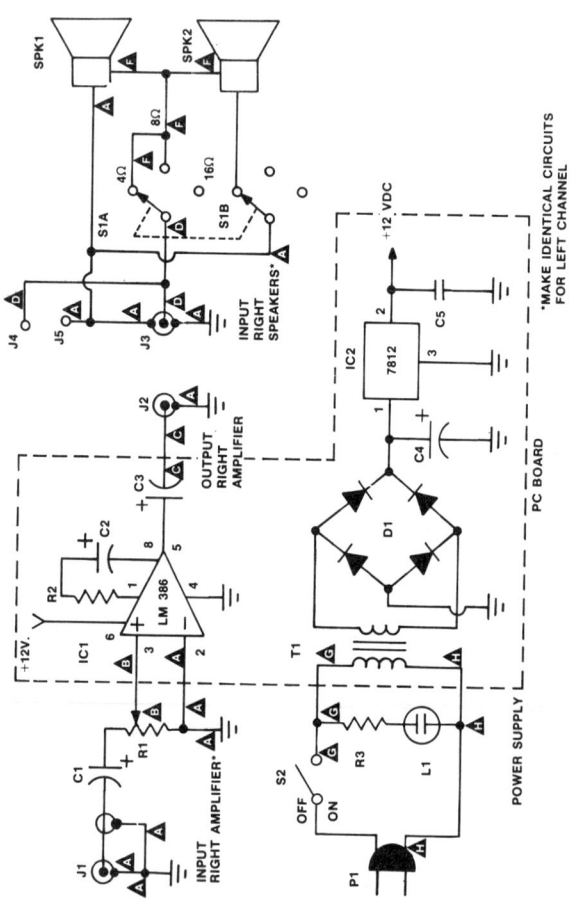

FIGURE 2—1A: AUDIO TEST JIG CIRCUIT

Parts to Use

*C1, C2 – 10 MF, 25V
*C3 – 250 MF, 25V
C4 – 1000 MF, 25V
C5 – .1 MF, 25V
D1 – 1A, 50 PIV bridge
*IC1 – LM 386 audio
   amplifier
*IC2 – 7812 regulator
*J1-J3 – phono jack
*J4, J5 – 5-way binding
   post

L1 – NE-2 neon lamp
P1 – AC plug
*R1 – 10K potentiometer,
   audio taper
*R2 – 1K ± 10% ½W
R3 – 56K ± 10% ½W
*S1 – 2-pole, 3 pos.
   rotary switch
S2 – SPST switch
*SPK1, SPK2 – 8 ohm speaker
T1 – 117:12 VAC, 300 MA
   transformer

Misc.– coaxial cable, enclosure, hardware, line cord,
   neon lens and clip, PC board, silicon grease,
   strain relief, terminal strips, etc.
*Need identical items for second channel

The PC Layout You'll Need

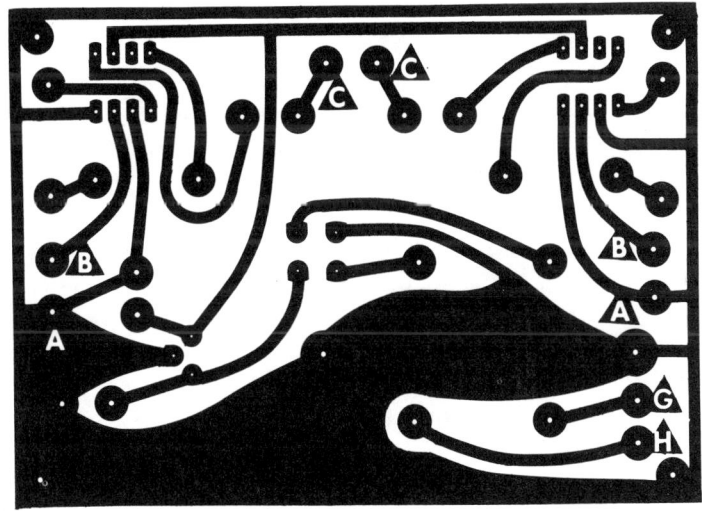

**FIGURE 2—1B: AUDIO TEST JIG PC LAYOUT (ACTUAL
                 SIZE)**

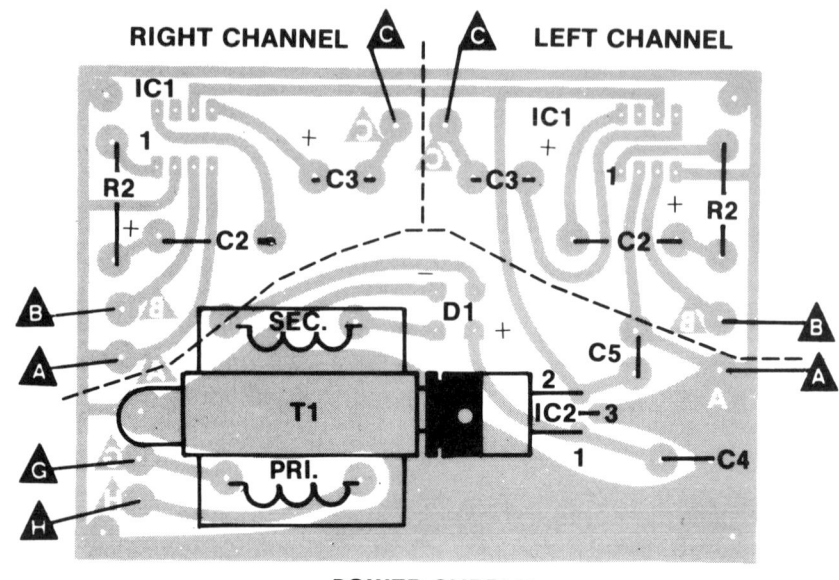

POWER SUPPLY

FIGURE 2—1C: AUDIO TEST JIG COMPONENT PLACE-
MENT GUIDE (TOP VIEW)

Helpful Construction Hints

Label the front panel with dry transfer labels as shown in Figure 2—1D. All controls, jacks and speakers are mounted on the front panel. The remaining parts are soldered on the printed circuit.

The 7812 voltage regulator IC operates fairly hot. Heat sinking it to the transformer frame using the transformer mounting screw (see Figure 2—1E), helps to dissipate the heat but some chassis ventilation would also be helpful. For good thermal conductivity, dab some silicon grease between the regulator tab, the transformer frame, and the PC foil before fastening them together.

Try to keep C1, potentiometer R1, and Build-O-Matic Guide identification point B on the PC board as physically close as possible to avoid hum and pickup problems.

Rubber feet and a carrying handle make a nice extra feature for increasing the portability of the test jig.

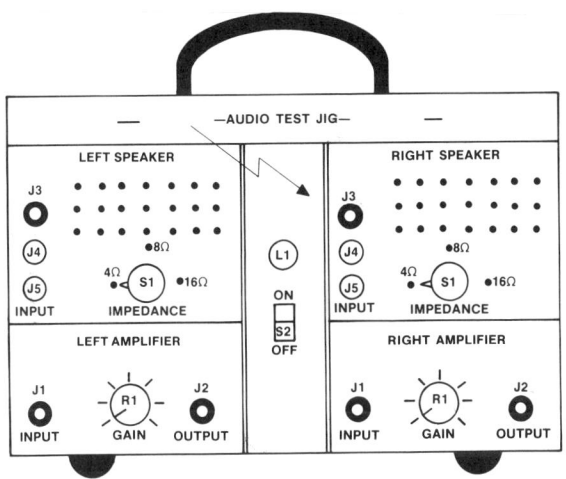

FIGURE 2—1D: AUDIO TEST JIG FRONT PANEL LAYOUT

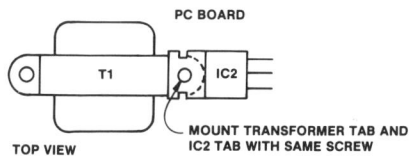

FIGURE 2—1E: MOUNTING IC2 TO THE TRANSFORMER FRAME

How It Works

The audio test jig is designed so that either the speakers and/or amplifiers can be used individually or together. S1 and its counterpart in the left channel select speaker impedances of 4, 8, or 16 ohms by switching the speaker connections in a series or parallel arrangement. If an amplifier is desired with a speaker system, simply patch the output of the amp to the input of its respective speakers with test leads terminated with phono plugs.

The heart of each amplifier is the LM 386, a one-watt audio amplifier IC. The gain of the input signal is controlled on each amplifier by potentiometer R1. Low cost and minimum parts count make the LM 386 ideal for use as utility amplifiers in the audio test jig.

A transformer-operated full wave power supply, regulated at twelve volts by the 7812 voltage regulator IC, provides the power for the amplifiers.

Operating Tips

Make an assortment of phono plug terminated test leads so that all combinations of the audio test jig will be easily available.

Operation is straightforward. On-off switch S2 does not have to be on for speaker only use. Just set the desired speaker impedances with S1 and plug into the appropriate speaker jacks.

Each amplifier can be used separately or with its speaker system depending on your needs. The choice is yours with just a few phono plug connections.

2—2: SIGNAL TRACER

What It Does

Here is a compact and inexpensive signal tracer that rivals the larger and more costly commercial units. This indispensable test instrument allows you to locate that defective stage in audio and radio equipment quickly and effectively. Incorporating a high impedance FET input and built-in detector, IF and even weak RF signals are tracked down with ease. In addition to high sensitivity, this tracer offers more than ample volume from its 2-watt IC amplifier. Add this handy instrument to your workbench equipment, and make your troubleshooting more fun and profitable.

Here's the Schematic

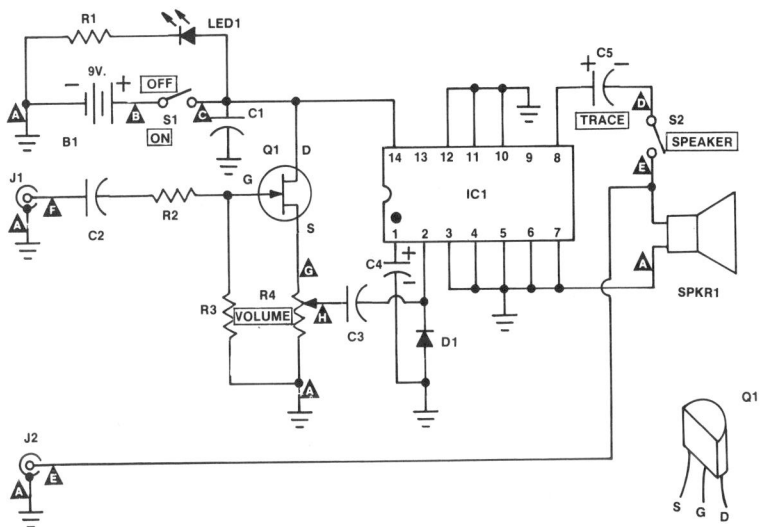

FIGURE 2—2A: SIGNAL TRACER

Parts to Use

B1 – 9 volt transistor battery	Q1 – junction field effect transistor, N-channel, 2N3819 (Radio Shack 276-2035)
C1, C3 – .1 mfd. 25 volt	
C2 – .22 mfd. 400 volt	
C4 – 47 mfd. 15 volt	R1 – 330 ohm, ¼ watt, 10%
C5 – 470 mfd. 15 volt	
D1 – germanium signal diode (1N34A, 1N60)	R2 – 47 K ohm, ¼ watt, 10%
J1, J2 – RCA phono jacks	R3 – 10 meg ohm, ¼ watt 10%
IC1 – LM380N, 2 watt amplifier	R4 – 5 K ohm audio taper
	S1, S2 – SPST switches
LED1 – jumbo red	SPKR1 – 8 ohm, 3-4″ PM speaker

Misc.- assorted hardware, battery connector, enclosure, hookup wire, IC socket (14 DIP, low profile), knob, PC board, etc.

The PC Layout You'll Need

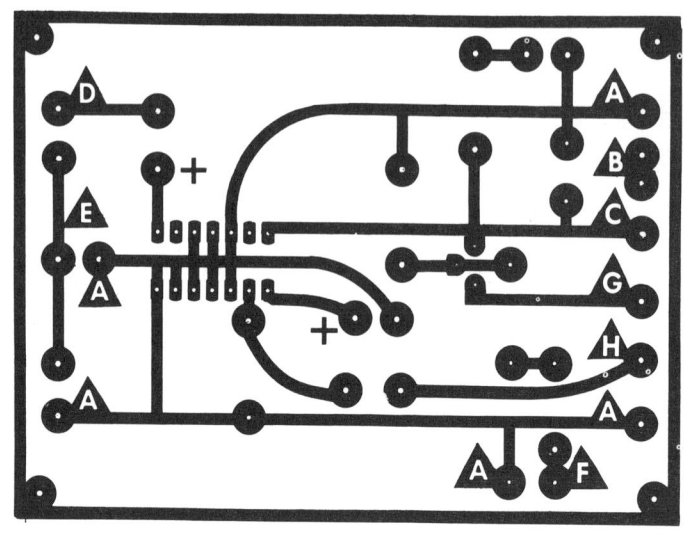

FIGURE 2—2B: SIGNAL TRACER PC LAYOUT (ACTUAL SIZE)

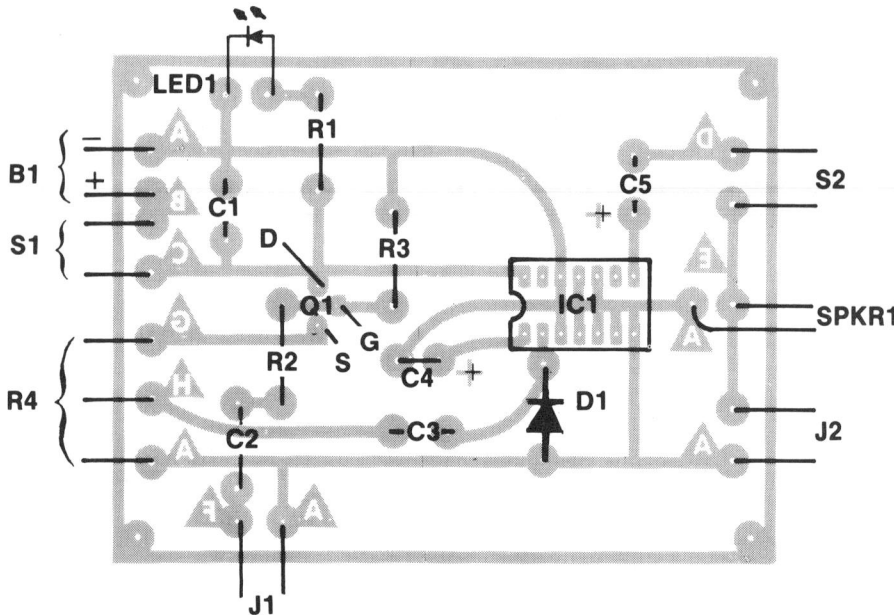

FIGURE 2—2C: SIGNAL TRACER COMPONENT PLACE-MENT GUIDE (TOP VIEW)

Helpful Construction Hints (Refer to Figure 2—2C)

Insert and solder the component parts in the printed circuit board, carefully noting the lead placement of Q1, and the polarities of the capacitors, diode, and the LED. A small soldering iron should be used, particularly on the IC socket pads, to prevent bridging and/or lifting of the copper-clad.

To minimize oscillation and feedback, the following construction techniques are recommended. Gently twist together the leads from R4 to the printed circuit board. Use shielded phono cable for wiring J1 to C2 and J2 to S2. Carefully route and separate the input and output leads.

When installing the IC in the socket, pay particular attention to the pin configuration. A small notch in the IC's case designates pins 1 and 14.

How It Works (See Figure 2—2A)

Closing S1 applies power to the LED, Q1 and IC1. C2 couples the input signal to the JFET, a high impedance voltage amplifier. R2 and R3 form a voltage divider input load for the gate of Q1. Functioning as a volume control, R4 determines the amount of output signal coupled to IC1, via C3. As the wiper arm of R4 is moved in the direction of ground, a decrease in signal tracer volume will result. Demodulation of RF and IF signals is accomplished by D1.

C1 suppresses amplifier oscillations that might otherwise enter the power source. IC1 signal bypass is accomplished by the electrolytic capacitor C4. Output coupling is provided by C5, a DC blocking capacitor.

S2 is normally closed in the signal tracing mode. However, opening S2 and using input J2 and ground, the substitute speaker function may be utilized.

Operating Tips

Plug the probe lead into J1. Close S1 and S2. Connect the tracer's ground lead to the chassis ground of the receiver or amplifier being tested. Beginning at the input, probe the input and output of each stage, progressing toward the unit's output. When a signal is no longer heard, the problem exists between that point and

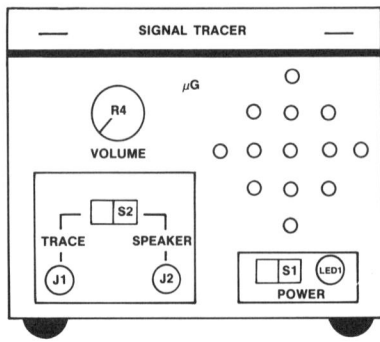

FIGURE 2—2D: SIGNAL TRACER (FRONT PANEL VIEW)

the previously probed area. Phono cartridges, microphones, and tape recorder heads can also be easily checked with the signal tracer.

Open S1 and S2. Plug the probe into J2. A handy substitute 8-ohm speaker is now available for experimenting or troubleshooting.

2—3: MULTIPURPOSE DC POWER SUPPLY

What It Does

This slick power supply is just the ticket for all sorts of applications. It's variable from 1.25 volts to 25 volts with a current capability of 1.5 amps. Not only is the supply protected from overload conditions and has a high ripple rejection, but it also has super line and load regulation.

The LM317 regulator IC, the heart of the supply, provides current limiting, power limiting and thermal shutdown protection. It's a mighty nice supply that's easy to make with minimum parts.

Here's the Schematic

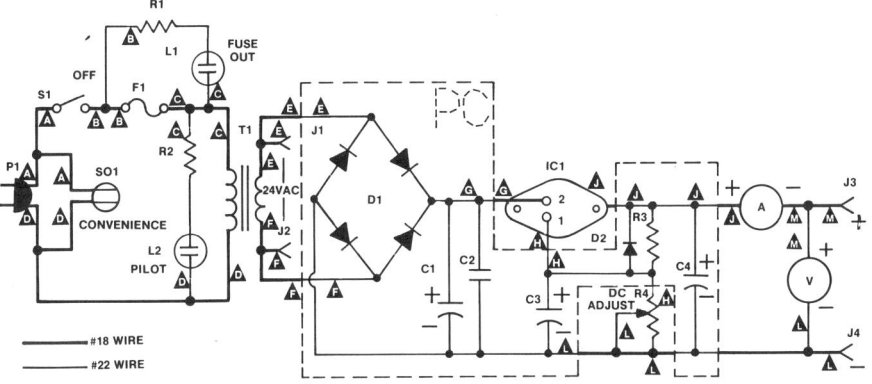

FIGURE 2—3A: MULTIPURPOSE DC POWER SUPPLY

Parts to Use

C1 – 2000 MF 50V	L1, L2 – NE-2 neon lamp
C2 – .1 MF, 50 V	M1 – 0-2A ammeter
C3 – 10 MF, 50 V	M2 – 0-25V voltmeter
C4 – 2.2 MF, 35 V tantalum	P1 – AC plug
D1 – 2A, 100 PIV bridge	R1, R2 – 56K ±10% ½W
D2 – ½A, 100 PIV	R3 – 270 ohm ±10% ½W
F1 – ½A fast blow	R4 – 5K linear taper
IC1 – LM317 regulator	S1 – SPST
J1, J2 – banana Jack,	SO1 – convenience outlet
yellow	T1 – 117V:24V, 2A
J3 – five way binding	
post, red	
J4 – five way binding	
post, black	

Misc.- enclosure, fuse holder, hardware, heat sink, knob, line cord, neon lens caps and clips, PC board, silicon grease, etc.

The PC Layout You'll Need

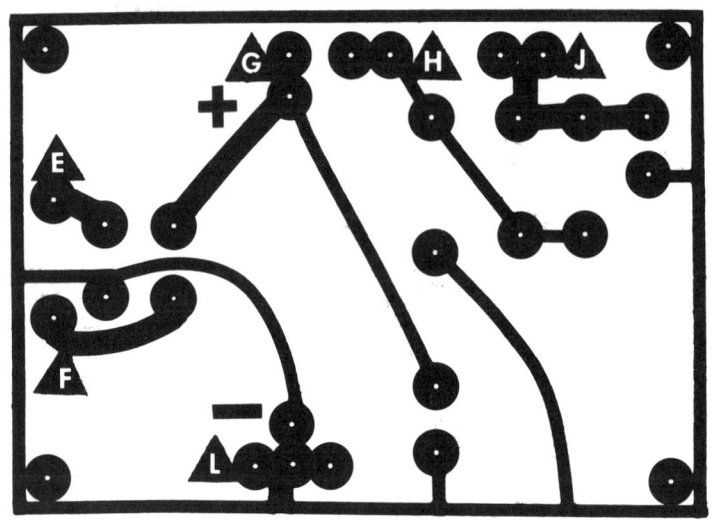

FIGURE 2—3B: MULTIPURPOSE DC POWER SUPPLY PC
LAYOUT (ACTUAL SIZE)

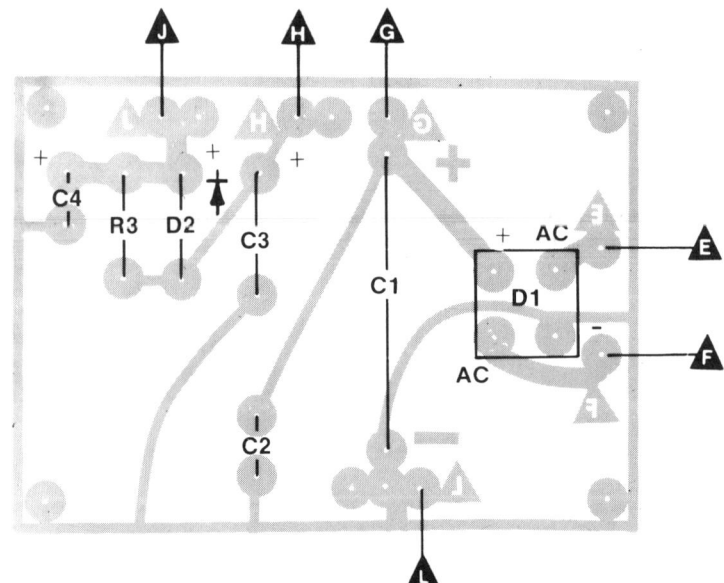

FIGURE 2—3C: MULTIPURPOSE DC POWER SUPPLY
COMPONENT PLACEMENT GUIDE (TOP
VIEW)

Helpful Construction Hints

Insulate the LM317 regulator IC from its heat sink with a mica washer and fiber shoulder washers as shown in Figure 2—3D. Use a heat sink that measures at least 3″ × 4″ × 1″ in order to dissipate the regulator heat. Coat both sides of the mica washer with silicon grease to insure good thermal conductivity.

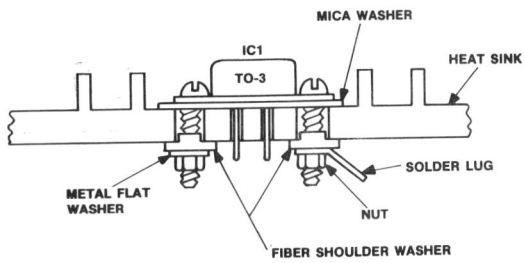

FIGURE 2—3D: INSULATING THE REGULATOR IC FROM ITS HEAT SINK

Use 18 gauge wire for the secondary circuit that carries the main power supply current (shown in heavy black lines on Figure 2—3A). The rest of the power supply can be wired with 22 gauge hookup wire. Physically mount the PC board as close to the regulator and output jacks as possible to provide optimum operation.

How It Works

The transformer (see Figure 2—3A) steps the line voltage down to 24 volts at the secondary. Two banana jacks are wired in parallel with the secondary winding to provide a convenient source of 24 VAC power. The bridge rectifier changes the AC to full wave DC. Filter capacitor C1 smoothes out the pulsating DC current.

R2 and potentiometer R4 forms a voltage divider for the adjustment terminal of the regulator IC (pin 1). C3 in parallel with the potentiometer filters out most of the remaining ripple voltage. Bypass capacitor C2 helps to maintain power supply stability. Output capacitor C4 also helps with stability as well as improving transient response and noise rejection. Rectifier D1 gives a discharge path for C3 if the output terminals are accidentally shorted. The two meters monitor the output current and voltage.

The LM 317 IC converts an ordinary run-of-the-mill supply into a very sophisticated bench power supply that should take care of most of your power supply requirements. It supplies excellent line and load regulation as well as providing current, power and thermal protection.

Operating Tips

A convenience outlet and 24 VAC jacks have been included in the supply for additional usefulness to this project. A blown fuse will be readily apparent because the "Fuse Out" neon will automatically turn on when the fuse becomes defective.

When the DC adjust pot is turned fully CCW, minimum regulated voltage (1.25 VDC) is available at the output jacks. CW rotation of the pot will give a uniform rise in voltage at the output until the maximum level is reached (25 VDC).

This is a rugged power supply and should stand up to a multitude of abuses that all bench supplies must take from time to time.

2—4: VARIABLE AC POWER SUPPLY

What It Does

A source of variable AC power is a must in the electronic workshop. This handy inexpensive supply provides 500 watts of AC power adjustable from a few volts to 100 volts.

For added flexibility, the variable AC voltage is available at an AC outlet and at a pair of banana jacks. The output power is monitored by a 7W lamp whose brightness magnitude gives a visual indication of relative voltage amplitude. Fixed line voltage is also available at a convenience outlet.

Not only will you use this unit as a bench supply, but also it's great for converting all your power hand tools into variable speed devices.

Here's the Schematic

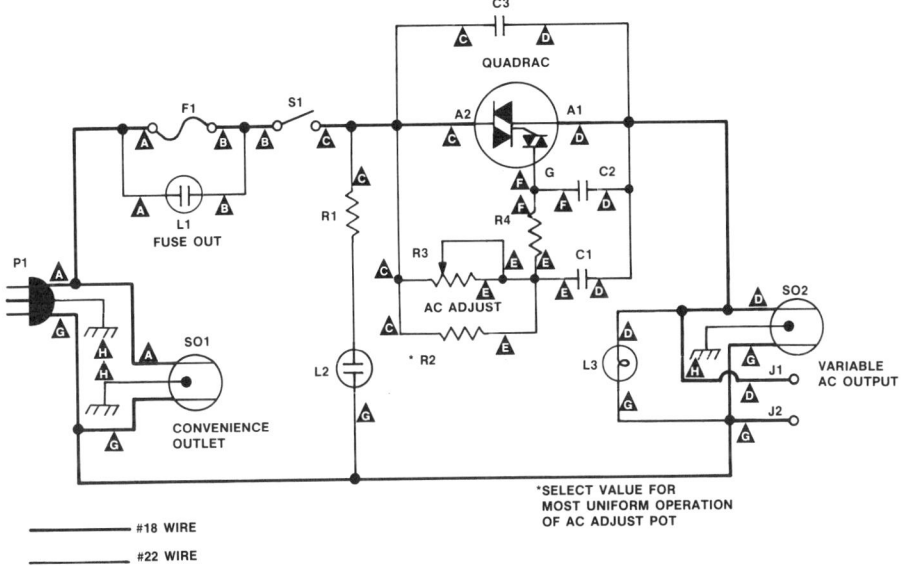

**FIGURE 2—4A: VARIABLE AC POWER SUPPLY
SCHEMATIC**

Parts to Use

C1, C2, C3 – .1MF disc ceramic 200V	Quadrac – 200V, 6A
	R1 – 33K ohms, ±10% ½W
F1 – 5A fast blow fuse	R2 – to be determined
J1, J2 – banana jacks, yellow	R3 – 100K ohm linear potentiometer
L1, L2 – NE-2 neon lamp	R4 – 22K ohms, ±10% ½W
L3 – 7W incandescent lamp	S1 – SPST switch
P1 – 117V AC plug	SO1, SO2 – 117V AC outlet
Misc.- fuse post, hardware, lamp socket, lens cap and clip, line cord, silicon grease, terminal strips, etc.	

Helpful Construction Hints

Most of the parts are mounted on the front panel, Figure 2—4B, and terminal strips so a PC board is not necessary for this project.

If the quadrac's heat sink tab is electrically isolated from the leads, it can be mounted directly to the metal enclosure. Be sure to insulate the heat sink from the chassis if the tab is not isolated from the leads. Be absolutely certain that no part of the circuit is in electrical contact with the chassis. Use silicon grease for maximum heat transfer from the quadrac to the heat sink.

Mount the 7W indicator lamp in an appropriate socket or you might want to use a large grommet to hold the lamp. Simply push the lamp partly through the grommet and solder the connections directly to the lamp's base. It makes a novel and neat looking job.

Use a resistor substitution box to find the value of R2. Clip the sub box leads across the two outside terminals of R3. Start at high resistance decreasing the resistance until the most linear potentiometer operation is obtained.

After the AC supply is operational, calibrate the AC adjust pot. Make a dial using dry transfer labels. Put a load on the power supply while making the calibration.

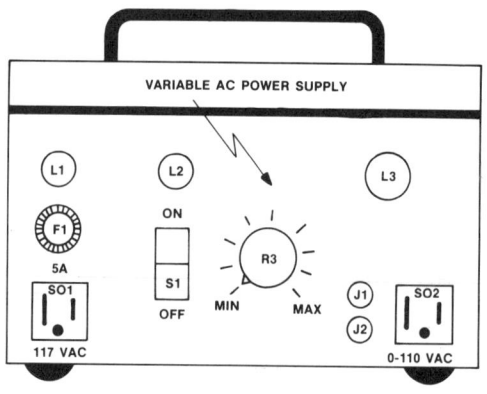

FIGURE 2—4B: FRONT PANEL LAYOUT

How It Works

The heart of this unit is a quadrac, a solid state bidirectional switch with built-in triggering circuitry. The output of the AC power supply (J1, J2 and SO2) is in series with the quadrac across the AC line. Please refer to Figure 2—4A. The setting of the AC adjust potentiometer determines how much of the AC cycle the quadrac will conduct, which in turn determines the power the load receives.

AC adjust pot, C1, R4 and C2 make up a phase-shifting circuit that determines the quadrac conduction interval. Rotation of R3 will provide a smooth continuous control of AC power. R2 across the AC adjust pot provides for more linear operation of R3.

C3 helps to reduce radio interference which is generated when the quadrac is turned on and off. The brightness of L3 shows relative power output. Neon lamp L1 acts as a blown fuse indicator when F1 opens. SO1 is included as a convenience outlet.

Operating Tips

The output of the AC supply is connected directly to the AC line so be certain to use proper safety precautions. Be sure to keep the output leads from shorting each other or the fuse will blow.

The operation of the supply is straightforward and should give you years of reliable service.

2—5: CAPACITOR TESTER

What It Does

This neat capacitor tester will check a wide variety of nonelectrolytic capacitors. All values from fifty picofarads to one microfarad can easily be checked for relative capacity, leakage, opens and shorts. Leakage as high as twenty million ohms can be detected.

The tester also can serve as a continuity tester for all sorts of electrical apparatus.

Here's the Schematic

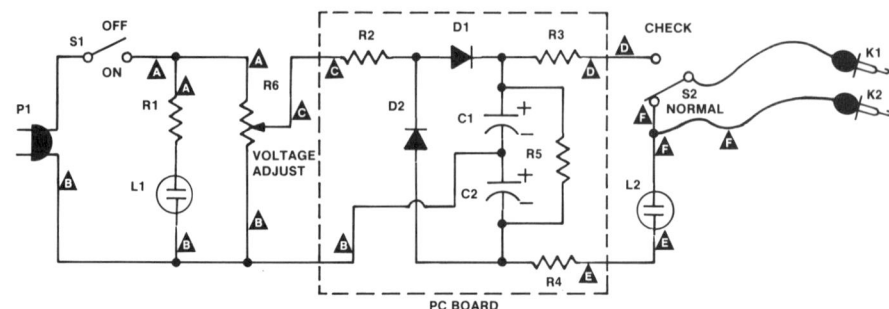

FIGURE 2—5A: CAPACITOR TESTER SCHEMATIC

Parts to Use

C1, C2 – 20MF, 150V electrolytic   R2 – 2.7K ±10% ½W
D1, D2 – ½A, 400 PIV   R3, R4 – 120K ±10% ½W
K1 – IC test clip, red   R5 – 1M ±10% ½W
K2 – IC test clip, black   R6 – 5K, 5W wire wound linear taper potentiometer
L1 – NE-2 neon
L2 – NE-51 neon   S1 – SPST switch
P1 – AC plug   S2 – SPDT momentary action switch
R1 – 47K ±10% ½W

Misc. – enclosure, hardware, knob, line cord, PC board, strain relief, test lead wire, etc.

The PC Layout You'll Need

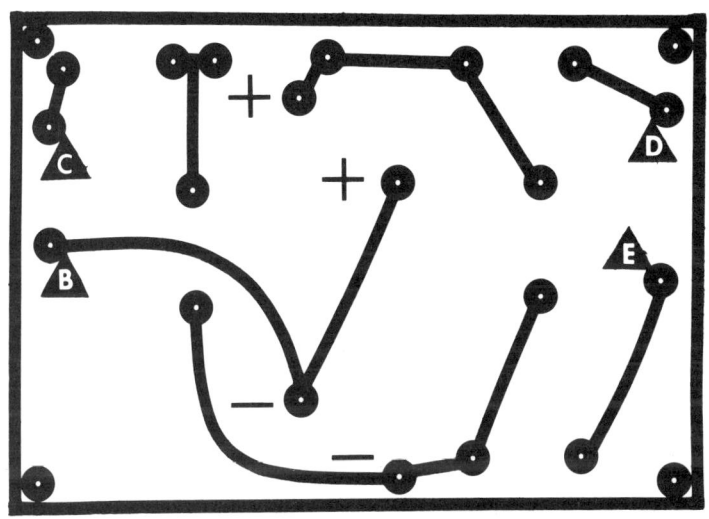

FIGURE 2—5B: CAPACITOR TESTER PC LAYOUT (ACTUAL SIZE)

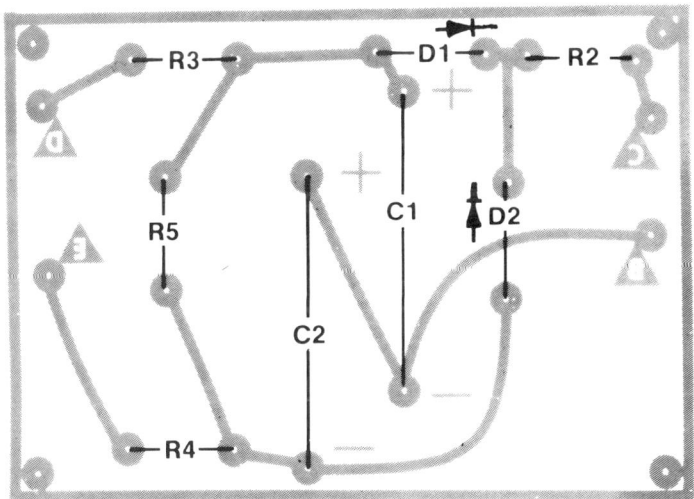

FIGURE 2—5C: CAPACITOR TESTER COMPONENT PLACEMENT GUIDE (TOP VIEW)

Helpful Construction Hints

Mount the PC board in a phenolic instrument case or a metal minibox being sure that it is electrically isolated from the box. Since this tester is line operated it is essential that the box (if metal) is insulated from all circuit parts.

Be sure to use an SPDT momentary action switch for S2. After testing the capacitor, the switch will automatically return to its normal position, shorting out the capacitor and thereby removing the charge. Calibrate the potentiometer (R6) by monitoring the output leads' voltage while testing a capacitor at various voltage settings from just a few volts to maximum voltage.

How It Works

Please refer to Figure 2—5A for the following explanation. Closing S1 applies power to the pilot light (L1) and potentiometer R6. R6 varies the input voltage from zero to full line voltage for the remainder of the circuit. D1, D2, C1, and C2 form a voltage doubler network. C1 will charge to peak voltage set by R6 on one-half of the AC cycle. C2 will do likewise on the other half of the AC cycle. Since both capacitors are in series across the output, their combined voltages will add. When S2 is placed in the check position, C1 and C2 voltage appears across the red and black IC test clips. The capacitor under test will charge up to the output voltage through the NE-51 neon lamp (L2).

Operating Tips

Close S1. Set R6 for the working voltage of the capacitor being checked. Connect the capacitor between the red and black IC test clips. Place S2 in the check position. A good capacitor will cause L2 to blink once as the capacitor charges to the output voltage. The duration of the blink is a factor of the capacity. A large capacitor will cause a relatively long blink. A small capacitor produces a very short blink. If the capacitor is leaky, L2 will blink at a rate determined by the amount of leakage; lots of leakage—a fast blink rate, little leakage—a slow blink rate. L2 will remain on for a shorted capacitor and remain off for an open capacitor.

2—6: HI-LO VOLTAGE PROBE

What It Does

You will find the hi-lo voltage probe, Figure 2—6A, an indispensable helper when checking for the presence of voltage. It will easily check voltage magnitudes from fifty volts to many thousands of volts. Not only does the probe check relative amplitude but it distinguishes AC from DC and determines DC polarity.

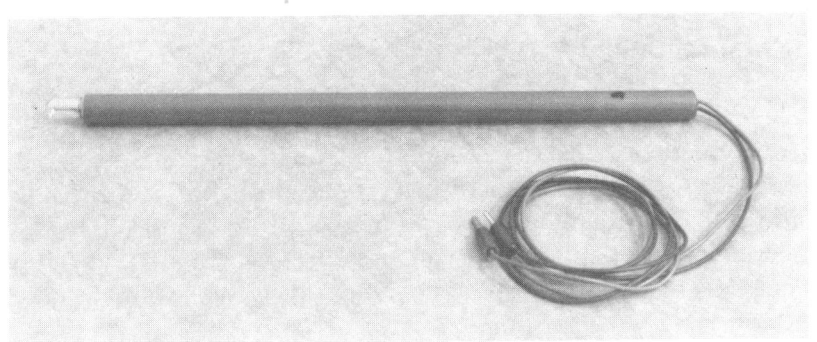

FIGURE 2—6A: HI-LO VOLTAGE PROBE

The hi-lo voltage probe is also ideal for troubleshooting electronic devices where high AC voltage fields are present, such as in a television receiver. The "hi" end of the probe will actually light up when placed in a high AC voltage field. Since there is no physical contact with the circuits there's no worry about loading or disturbing the section being tested.

Here's the Schematic

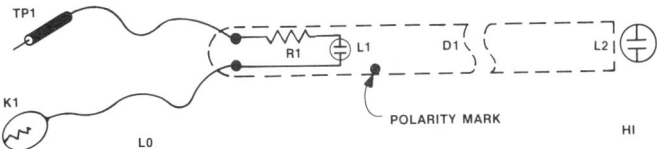

FIGURE 2—6B: HI-LO VOLTAGE PROBE SCHEMATIC

Parts to Use

D1 – ½" × 1' dowel rod,   L2 – NE-51 neon lamp
      wood or clear plastic
K1 – clip-minigator,   R1 – 47K ±10% ½W
      black insulated
L1 – NE-2 neon lamp   TP1 – test prod, red
Misc. – epoxy glue, strain relief bushing, 2' test lead wire
      (red), 2' lead wire (black), etc.

Helpful Construction Hints

Drill a ⅜" hole in each end of the dowel rod as shown in Figure 2—6C. A lathe is ideal for this operation but if one is not available, a satisfactory job can be done with a hand drill and vise. If using a hand drill start the end holes with a ⅛" drill bit, centering them as accurately as possible. Work your way up to the final ⅜" size with a succession of drill sizes. This technique will give you enough control for good alignment of the end holes.

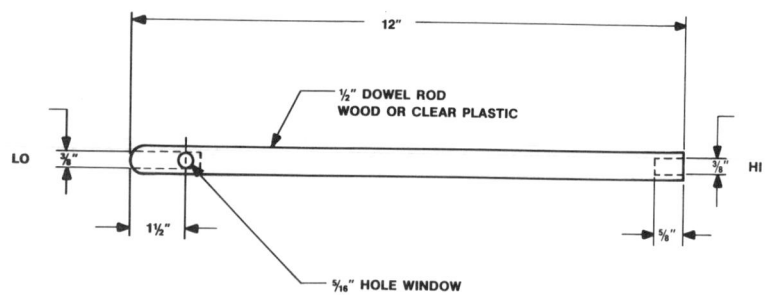

FIGURE 2—6C: PROBE BODY CONSTRUCTION

The 5/16" hole window is necessary to see the NE-2 lamp which is mounted entirely inside of the dowel. Position the NE-2 so the electrodes are visible through the window. The darkness of the probe body will enhance the NE-2 illumination when making checks with the test leads.

The NE-51 can be held in place with a drop of epoxy glue. File off the bayonet base tips so that it will slide easily into the probe end. Position the resistor and NE-2 assembly so that it is in its proper place when connected to the test leads. Pop a strain relief bushing on the test leads and push the assembly into the probe body.

How It Works

When using the test leads to measure voltage, the NE-2 neon gas will ionize for voltages greater than 50 volts. The resistor is absolutely necessary to limit the NE-2 current. The relative brightness of the NE-2 is a function of the voltage amplitude. AC voltage will cause the gas around both electrodes to glow. DC voltage will cause the gas around just one electrode to glow. The electrode that is connected to the most negative test lead will be the one where the gas is ionized.

Operating Tips

It's a good idea to place a mark on the probe body close to the NE-2 window so that you know which test lead wire is connected to each of the NE-2 electrodes. (See Figure 2—6B.) This identification mark will enable you to easily determine the most negative test lead when checking the polarity of a DC circuit.

Needless to say, good safety practices should be observed when checking for the presence of high AC voltage. The probe body is made twelve inches long so that you can safely stay away from high voltage sources. Be sure you are standing on an insulated surface and keep one hand behind your back, or in your pocket when performing a high voltage test.

A good example of high AC voltage fields is in a television receiver. An operating TV will have AC voltage fields present around the horizontal output section, flyback transformer, damper, and high voltage rectifier sections. Checking for the presence or absence of these fields gives the troubleshooter valuable clues for finding the problem area.

2—7: MICRO SIGNAL INJECTOR

What It Does

Here's a troubleshooting tool (Figure 2—7A) that can't be beat for fast, efficient repair work on all sorts of electronic devices. You will really appreciate its small size and simplicity of operation.

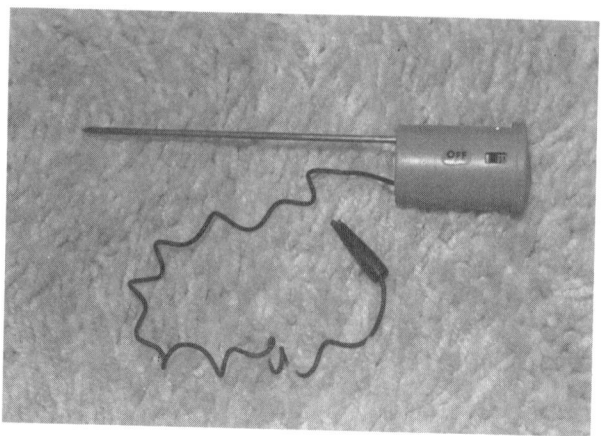

FIGURE 2—7A: MICRO SIGNAL INJECTOR

A signal injector is a miniature audio/radio transmitter that can be used to inject test signals for localizing trouble in radios, tape recorders, hi-fi systems, CB's, televisions, and other electronic apparatus. Many electronic problems are caused by a defect that blocks or distorts the normal operation of the device. The signal injector makes it possible to easily feed a test signal anywhere in the signal path zeroing in on the problem area.

Here's the Schematic

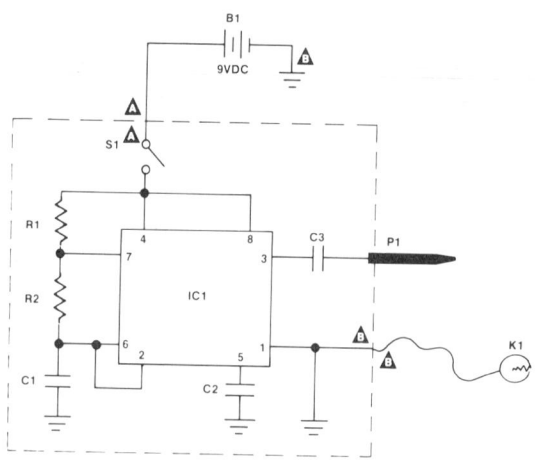

FIGURE 2—7B: MICRO SIGNAL INJECTOR CIRCUIT

Parts to Use

B1, battery, 9V
C1, C2 - .01 MF, 15 VDC   P1 - ⅛" × 6" brass rod
C3 - .01 MF, 200 VDC   R1, R2 - 100 K ±10% ¼W
IC1 - 555 timer   S1 - SPST miniature
K1 - clip/minigator,   slide switch
    black insulated
Misc. – battery connector (9V), film case, fish paper insula-
    tion, PC board, 2' test lead wire (black), etc.

The PC Layout You'll Need

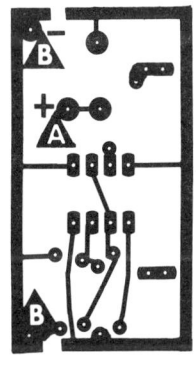 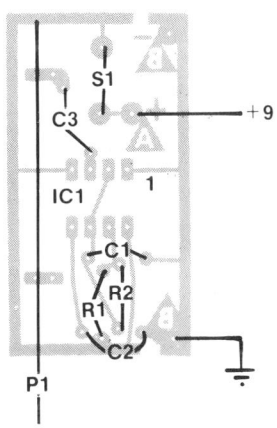

**FIGURE 2—7C: MICRO SIGNAL INJECTOR PC LAYOUT
                (ACTUAL SIZE)**
**FIGURE 2—7D: MICRO SIGNAL INJECTOR COMPONENT
                PLACEMENT GUIDE (TOP VIEW)**

Helpful Construction Hints

The entire signal injector plus battery is designed to fit into a 35 MM film case. The three capacitors are the only parts that must be selected for minimum physical size.

Build the prod from a ⅛" brass rod six inches in length. File one end of the rod to a point. The other end of the rod can be soldered directly to the PC board. Drill a hole in the bottom of the film case exactly at the spot where the brass rod will exit when the PC board is slid into the case.

The slide switch should also be soldered directly to the PC board. In this way all parts of the signal injector are an integral part of the PC, simplifying construction. Carefully drill two holes in the film case where the switch lever will be located. A small square file can then be used to shape the small rectangular opening necessary for the switch lever.

Mark the "off" position on the film case so the battery is off when not being used. When ready for final assembly, insert the PC assembly into the case first. Position a strip of fish paper insulation that is the same size as the PC board between the PC and the battery so the battery case will not short the bottom of the PC board. If the probe and switch holes are located properly there will be enough room to slide a 9V battery alongside of the PC. Pop the battery connector on, snap the film case lid shut and you are ready to go.

How It Works

The heart of the micro signal injector is the 555 timer IC. It is connected as an astable multivibrator. The free running frequency is determined by the two 100 K resistors and the .01 MF capacitor tied to pins 6 and 2 of the IC (Figure 2—7B).

The output waveform is a rectangular wave as shown in Figure 2—7E. Although the basic pulse rate is in the audio frequency range, the fast moving leading and trailing edges produce radio frequency harmonics suitable for driving a signal through a wide assortment of IF and RF amplifiers.

Operating Tips

Signal injection is usually started at the output of the apparatus and worked back towards the input along the signal flow path. When the defective stage is encountered, the output indication will be missing or distorted. As an example, suppose you are troubleshooting a bad tape player. Injecting a signal at the input to the final power amplifier would be a good place to start. If the injector tone is heard in the tape player's speaker you could assume that the power amplifier and the speaker are operating normally. Next, inject at the input to the preceding stage, probably a driver amplifier. The speaker tone should be heard again but with more

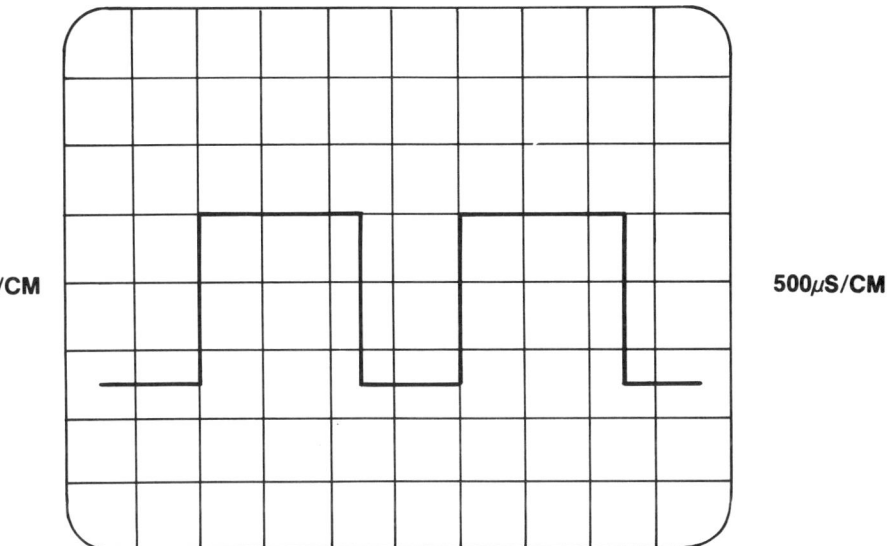

2V/CM

500μS/CM

**FIGURE 2—7E: MICRO SIGNAL INJECTOR OUTPUT
WAVEFORM**

volume if this stage is good. No sound or distortion would indicate
that the problem lies in the driver amplifier stage. If the driver is
okay, the signal injection process is continued systematically along
the signal path to the tape head until the malfunctioning area is
located.

2—8: TV TUNER HELPER

What It Does

Pinpointing a defective TV tuner can often be a frustrating
problem. Is the tuner really good or bad? This TV tuner substitution
unit really helps to remove the question mark from tuner problem
uncertainties.

Here's all you do. Simply disconnect the antenna input and the
IF output from the suspected band tuner. Reconnect these leads to
the tuner substitution unit. Presto! The TV is now operating from the
tuner substitution unit. It's that simple.

Since the tuner helper unit has its own battery supply for both
power, AFT, and AGC, the substitution procedure is fast and easy.

If the TV still does not work after substitution, you know the original problem is definitely not in the tuner. If the TV operation returns to normal, the defect is in the original tuner.

Here's the Schematic

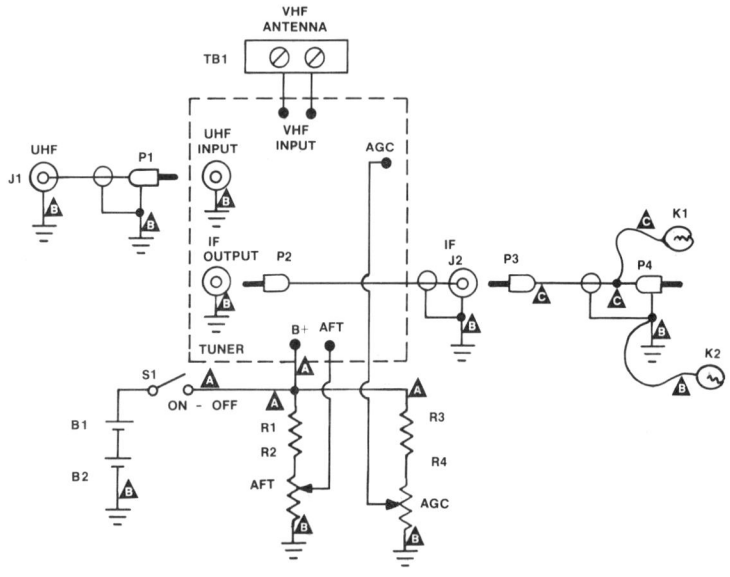

FIGURE 2—8A: TV TUNER HELPER SCHEMATIC

Parts to Use

B1, B2 - 9V alkaline battery	R1, R3 - 3.3K ohms ½W ±10%
J1, J2 – phono jack	R2, R4 - 1K ohm linear taper potentiometer
K1 – red insulated mini- gator clip	
	S1 - SPST switch
K2 – black insulated minigator clip	TB1 - 2 screw terminal board
P1-P4 – phono plug	Tuner - Sylvania #54-43651-2 or equivalent
Misc. - battery holders, coaxial cable (RG-58 A/U), enclosure, hardware, terminal strip, twin lead, etc.	

Helpful Construction Hints

The authors installed the tuner and associated components in an old VOM bakelite case. (See Figure 2—8B.) It's a ready-made enclosure that's a perfect size for this project. Drill holes for the tuner shaft, AFT and AGC potentiometers, on-off switch, IF output and UHF phono jacks, the antenna terminals, and mounting hardware. Everything else can easily fit inside of the enclosure.

FIGURE 2—8B: TV TUNER HELPER UNIT

A clip is mounted on the top cover plate to neatly hold the IF output coaxial cable when it's not being used. The IF output cable is terminated in both a phono plug and a parallel set of insulated minigator clips for odd hookup arrangements.

Use two 9V transistor radio batteries connected in series for the power, AFT, and AGC voltage. Alkaline batteries will last for months as long as the on-off switch is turned off after every test.

How It Works

The Sylvania VHF 54-43651-2 is a standard color tuner that normally operates on a 24 VDC power supply. The two series 9V batteries are adequately suited to supply the tuner, making it independent from the TV receiver. (See Figure 2—8A.)

For optimum fine tuning, AFT voltage is selected by clockwise rotation of the AFT pot. Manual fine tuning can also be made with the conventional fine tuning control.

Tuner AGC voltage is selected by clockwise rotation of the AGC pot. Just turn the pot until the best picture and sound are being received. Voltage divider networks R1, R2, and R3, R4, provide the proper AFT and AGC bias.

VHF and UHF signals are processed exactly the same as they would be in the original tuner. You are simply bypassing the original VHF tuner with the substitution tuner unit to help localize possible tuner problems.

Operating Tips

VHF signals are fed directly into the substitute tuner's VHF input terminals. All VHF stations in your area should work with the substitute tuner. However AFT and AGC voltage will probably have to be changed slightly for each channel.

UHF stations can also be received by removing the UHF output cable from the original tuner and plugging it into the substitute tuner's UHF input jack. Turn the substitute tuner's and original tuner's VHF channel selector knob to the UHF position. Remember to connect a UHF aerial to the TV's UHF antenna terminals. The remainder of the connections are the same as for VHF reception.

The substitute tuner may not always work as well as the original tuner for all TV models. This is as expected since every TV is not exactly alike. In almost all cases the reception will be more than adequate to help you diagnose the problem.

3

Energy-Saving Projects You Can Build

3

3—1: EMERGENCY FIREPLACE BLOWER

What It Does

With the high cost of energy, millions of families have equipped their fireplaces with energy efficient heat exchangers. Many of these heat exchangers operate in conjunction with a fan driven by an electric motor. The fan forces room air through the heat exchanger. The air is heated and returned to the room, increasing the useable fireplace heat output many times. The system works. It provides added warmth and reduces fuel bills.

There's only one catch. What happens when there is an electricity failure in the dead of winter? Most furnaces will not work without electric power. You can still use your fireplace, but most of the heat will be lost up the chimney. If the fireplace is equipped with a blower-operated heat exchanger, it will become most inefficient just when you need the heat. In fact, many heat exchangers require continuous forced air through them or they will overheat and burn out. Here's where the emergency fireplace blower-motor saves the day. Simply interchange the existing heat exchanger blower-motor with the battery-operated emergency fireplace blower unit. The emergency blower is designed to work off your car battery and will provide reliable service for hours.

Here's the Schematic

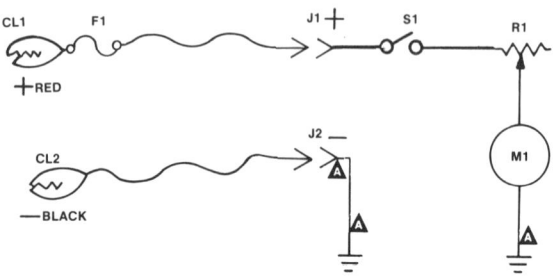

FIGURE 3—1A: EMERGENCY FIREPLACE BLOWER CIRCUIT

Parts to Use

CL 1 - Battery clip, red insulated	J2 - 5-way binding post, black
CL2 - Battery clip, black insulated	M1 - 12V automobile blower-motor
F1 - 5A slow blow fuse	R1 - 4 ohm, 20W potentiometer
J1 - 5-way binding post, red	S1 - SPST switch, 5A DC

Misc. – enclosure, hardware, in-line fuse holder, insulated wire—12 gauge, etc.

Helpful Construction Hints

An automotive heater blower-motor is ideal for the heart of this system. Choose one that is about the same physical size as the one in your heat exchanger. Most auto parts stores stock a variety of sizes. Another good source, and one that is much less expensive, is an automobile junk yard. You'll probably find just what you need lying on a shelf in the back corner of the shop. Pick a blower-motor assembly (see Figure 3—1B) that is about the right size and looks like it is in good condition. Spin the blower to see if it runs smoothly. Most junk yards will hook it up to a car battery so you can test its operation before you buy it.

FIGURE 3—1B: 12V BLOWER-MOTOR ASSEMBLY

Once you have your 12V blower-motor, decide how you can install it in the heat exchanger ductwork. The authors constructed a sheet metal box that is the identical size as the original blower-motor enclosure. (Refer to Figure 3—1C.) The blower-motor position is determined so maximum air is forced into the heat exchanger. Make sure the blower is turning in the correct direction to blow air *into* the ductwork and not *out*.

After the blower-motor is installed into the new enclosure, fit it to the ductwork in a similar way as the original. (Refer to Figure 3—1D.) Usually a few sheet metal screws will do the trick. Make it easy on yourself by designing the emergency blower unit so that it can be substituted easily and quickly for the original one. You won't want any delays in interchanging the units in an emergency situation.

FIGURE 3—1C: 12V BLOWER-MOTOR ENCLOSURE

FIGURE 3—1D: FINISHED EMERGENCY FIREPLACE BLOWER INSTALLED INTO EXISTING HEAT EXCHANGER

How It Works

Your car battery serves as the energy source for the emergency fireplace blower. Simply connect an appropriate length of 12 gauge insulated wire between the negative binding post of the fireplace blower (see Figure 3—1A) and the negative terminal of the car battery. Hook another 12 gauge insulated wire from the blower's positive binding post to the battery's positive terminal. Flick the on-off switch, adjust R1 for desired speed and heat should be pouring out of your heat exchanger.

Parking the car next to the house or in an attached garage is recommended to keep the connecting wires as short as possible. Figure out a system ahead of time to route the wires. You may want to permanently install the wires so that everything will be in place when an emergency arises.

Be sure to use an in-line fuse holder between the positive battery clip and the connecting wire to protect the system from accidental short circuits. When hooking up the system save the battery connections for last. Before clipping on to the battery terminals check to make sure all connections are good and that the on-off switch on the emergency fireplace blower is in the off position. This procedure will eliminate any chance of a spark occurring when the clips are attached to the battery terminals.

Operating Tips

The major concern in operating the emergency fireplace blower is the drain on the car battery. You won't want to run down the battery so much that the car won't start. Most blower-motors draw about 2 amps of current when they are running at a moderate speed. A car battery in good condition should be able to supply this amount of power for a day or night and still have enough reserve energy to start the car so the battery can be recharged.

It's impossible to give a hard and fast rule for the amount of time that the battery can supply the blower-motor. Variables such as type of blower-motor, kind of battery, battery condition, battery temperature, distance between battery and heat exchanger all play an important part in determining battery drain. Just play it on the safe side until you know the peculiarities of your system.

3—2: HEAT DUCT BOOSTER FAN

What It Does

If you would like to make your gas furnace more efficient and heat your home more evenly, build and install the heat duct booster fan. Mounted inside your existing ductwork, the booster fan increases the airflow to those difficult-to-heat areas of your home. By incorporating a relay control unit, even gravity type furnaces enjoy the advantages of a forced air system, without the costly changeover. Why wait? Save yourself money and increase your comfort by building and installing this energy-saving project.

Here's the Schematic

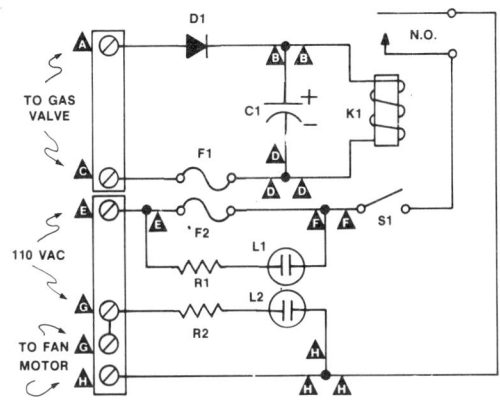

**FIGURE 3—2A: HEAT DUCT BOOSTER FAN RELAY
CONTROL UNIT**

Parts to Use

C1 - 100 mfd. 50 volt	K1 - 24 VDC SPST relay
D1 - 1 a. 50 PIV	L1, L2 - NE-2 neon lights
F1 - ½ a. fuse	R1, R2 – 56 K ohm, ½ watt 10%
F2 - 1 a. fuse	S1 - SPST switch

Misc.- assorted hardware, barrier terminal strips, electrical workbox and cover plate, enclosure, fuse holders, grommets, insulationg loom, hookup wire, 110 VAC shaded pole motor with fan, etc.

FIGURE 3—2B: RELAY CONTROL ENCLOSURE

Helpful Construction Hints

Mount the component parts in a minibox, carefully observing the polarities of D1 and C1. Point-to-point wiring is recommended, with an externally mounted terminal strip provided for connection to the gas valve. To expedite later installation, fasten an electrical workbox cover plate to the base of the enclosure and furnish a suitable terminal block for the line and motor wiring, as shown in Figure 3—2B. All wires passing through the cover plate and the enclosure should be protected with grommets.

How It Works (Refer to Figure 3—2A)

When the gas valve is actuated by the thermostat, the coil of K1 is energized. The closed relay contacts now supply power to the motor assembly, as indicated by the illumination of L2. To permit the use of a more readily available DC relay, D1 and C1 respectively rectify and filter the AC voltage applied to K1. F1 and F2 provide circuit protection, with L1 serving as an "open" fuse monitor. Manual fan shut-off is accomplished by S1.

FIGURE 3—2C: HEAT DUCT BOOSTER FAN

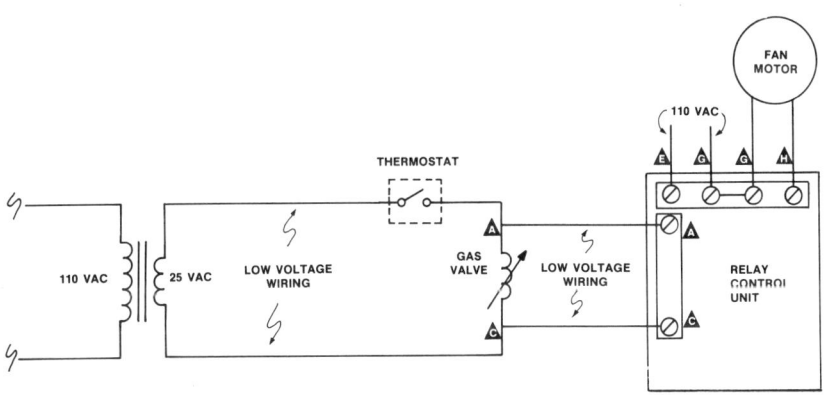

**FIGURE 3—2D: RELAY CONTROL UNIT CONNECTED TO
EXISTING THERMOSTAT CIRCUIT**

Operating Tips

INSTALLATION: Mount the motor/fan assembly inside the heat duct, using small angled brackets attached to the motor frame. (See Figure 3—2C.) Be sure the fan is near the heat register and oriented to draw air from the furnace. Next, furnish line, motor, and low voltage control wiring to an electrical workbox mounted near the heat duct. (See Figure 3—2D.) When installing the 110 VAC supply line, compliance with the local electrical code is recommended. The motor leads and the unprotected portions of the low voltage wiring should be covered with insulating "loom" for added safety. Complete the installation by wiring the relay unit into the circuit and securing the assembly to the workbox.

Note; Forced air heating systems do not require the relay control unit. Therefore, the booster fan may be wired directly to the existing blower-motor.

OPERATION: Close S1 for automatic control. When the furnace is activated, L2 will light, indicating fan operation. L1 is an "open" line fuse monitor, and is normally not illuminated. Open S1 to disable the fan.

Note; Installation of the heat duct booster fan may alter the airflow distribution. If necessary, readjust the duct dampers to ensure maximum heating efficiency and comfort.

3—3: HOT WATER TANK TIMER

What It Does

If you have an electric hot water tank, you can save up to 50% on the cost of water heating with the programmable hot water tank timer. Set the timer to provide hot water when you need it, and turn off the tank when you don't. Why spend money for something you're not using? Install the hot water tank timer and enjoy the savings.

Here's the Schematic

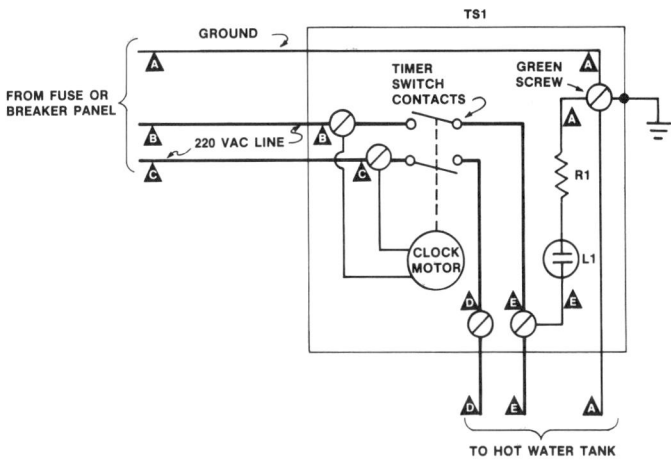

FIGURE 3—3A: HOT WATER TANK TIMER

Parts to Use

L1 – NE-2 neon light
R1 – 56 K, ohm, ½ watt, 10%
TS1 – 24 hour time switch (Intermatic T104)
Misc. - electrical fittings, hardware, lamp lens, terminal
strip, wire, etc.

Optional—additional trippers (if required)

Helpful Construction Hints

Mount the lens for L1 in the timer enclosure and provide a terminal strip for the neon and current limiting resistor. (See Figure 3—3A.) Be sure that the circuitry is electrically isolated from the timer enclosure.

Only the recommended timer or a comparable unit should be used. Inexpensive appliance timers are not suitable because of their 110 VAC clock motor mechanisms and low current switching capability.

How It Works (Refer to Figure 3—3A)

At preselected intervals, power is supplied to the water tank through a set of heavy duty switch contacts. These contacts are simultaneously actuated by the clock motor through a mechanical tripping mechanism. Separate, adjustable trippers are provided for programming the "on" and "off" modes of operation. Illumination of L1 indicates that power is available to the hot water tank.

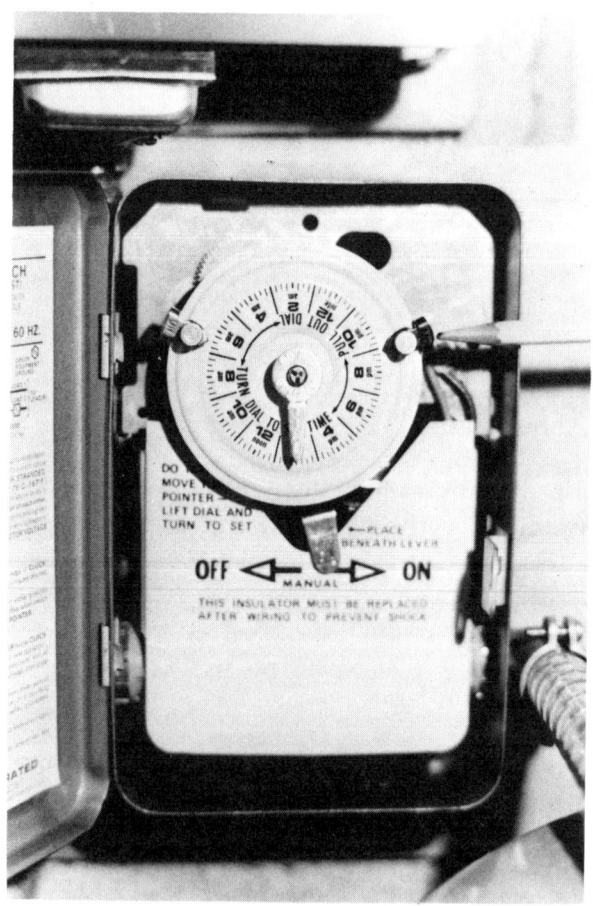

FIGURE 3—3B: TIMER PROGRAMMING

Installation (See Figure 3—3A)

Locate and mount the timer unit near the water heater, allowing easy access for programming and viewing of the indicator light. Before wiring the timer, turn off power to the hot water tank by removing both line fuses or tripping the circuit breakers to the "off" position. For added safety, verify that power is not being supplied with the aid of a voltmeter or neon test light.

Next, connect line and tank wiring to guide letters, B, C and D, E respectively. Fuse panel and tank ground connection are made at the green-colored screw (Guide letter A) attached to the timer backing plate. Wire comparable to that used in the original installation should be used. Be certain that all wiring conforms to the local electrical code.

Operating Tips

Determine the required on and off cycles of timer operation by completing the chart shown in Figure 3—3C, and set the trippers accordingly. If additional hot water is needed for clothes washing, bathing, etc. during the off timer cycle, use the manual override control.

To realize additional savings, and prolong the life of the water tank, the temperature may be reduced to 140 degrees (minimum temperature required for automatic dishwashers). By removing the front cover plate, access to an adjustable thermostat is possible. (See Figure 3—3D.) Adjustment of two thermostat controls is necessary if the water tank incorporates both upper and lower heating elements.

TIME	Major Hot Water Consumers		
	Clothes washing	Dishwashing	Baths & Showers
12:00 Midnight			
1:00 AM			
2:00 AM			
3:00 AM			
4:00 AM			
5:00 AM			
6:00 AM			
7:00 AM			
8:00 AM			
9:00 AM			
10:00 AM			
11:00 AM			
12:00 Noon			
1:00 PM			
2:00 PM			
3:00 PM			
4:00 PM			
5:00 PM			
6:00 PM			
7:00 PM			
8:00 PM			
9:00 PM			
10:00 PM			
11:00 PM			

DIRECTIONS: Mark an X at each time that a large quantity of hot water is required. It is most economical to arrange your hot water consumption so that it occurs within a one, two, or three-hour period each 24 hours. Careful planning of hot water usage pays off in dollars and cents.

FIGURE 3—3C: HOT WATER USAGE CHART

**FIGURE 3—3D: ADJUSTMENT OF WATER HEATER
THERMOSTAT**

3—4: TEMPERATURE ACTUATED VENTILATION
SYSTEM

What It Does

Here is a convenient energy saver for cooling your home. With the temperature actuated ventilation system, you can program your window or air exhaust fans to operate automatically when the room temperature exceeds a preset limit. Save money by using fans only when they are needed. Why not enjoy the comfort of a cooler and more energy-efficient home by installing the temperature actuated ventilation system?

Here's the Schematic

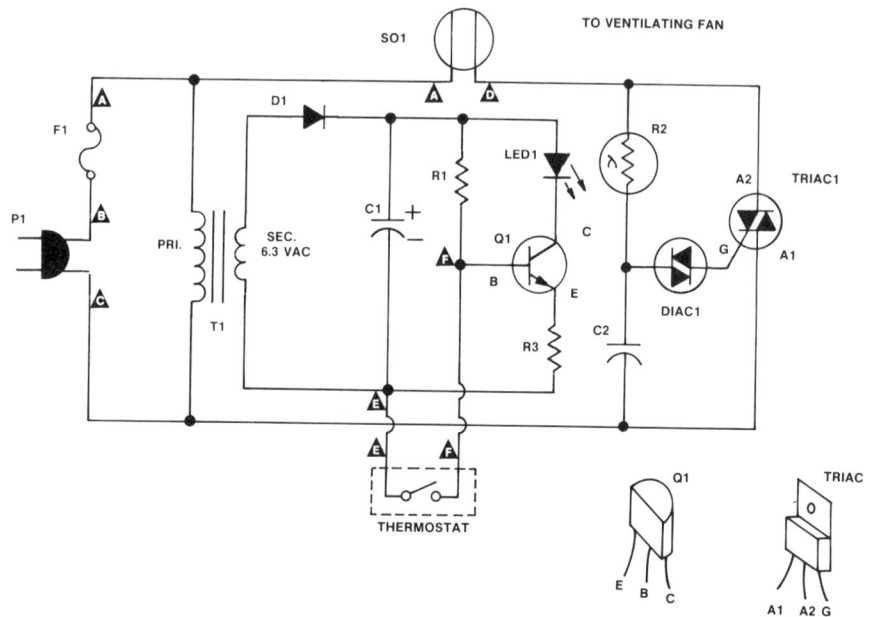

**FIGURE 3—4A: TEMPERATURE ACTUATED VENTILA-
TION SYSTEM**

Parts to Use

C1 – 470 mfd, 15 volt	F1 – 4 a. slow blow
C2 – .22 mfd, 200 volt	LED1 – jumbo clear
D1 – 1 a. 50 PIV	P1 – 110 VAC plug
DIAC1 – trigger diac, 28-32 volt (Radio Shack 276-1050)	Q1 – NPN low power transistor (2N3904 or equivalent)

R1 - 10 K, ohm, ½ watt, 10%   SO1 – 110 VAC outlet
R2 – cadmium sulphide   T1 – 6.3 volt, sec. @ 300 ma.
   photoresistor   TRIAC 1 – 6 a. 200 volt
   (Radio Shack 276-116)   thermostat – see text
R3 – 220 ohm, ½, 10%

Misc. - enclosure, fuse holder, grommet, hardware, hookup wire, line cord, PC board, etc.

The PC Layout You'll Need

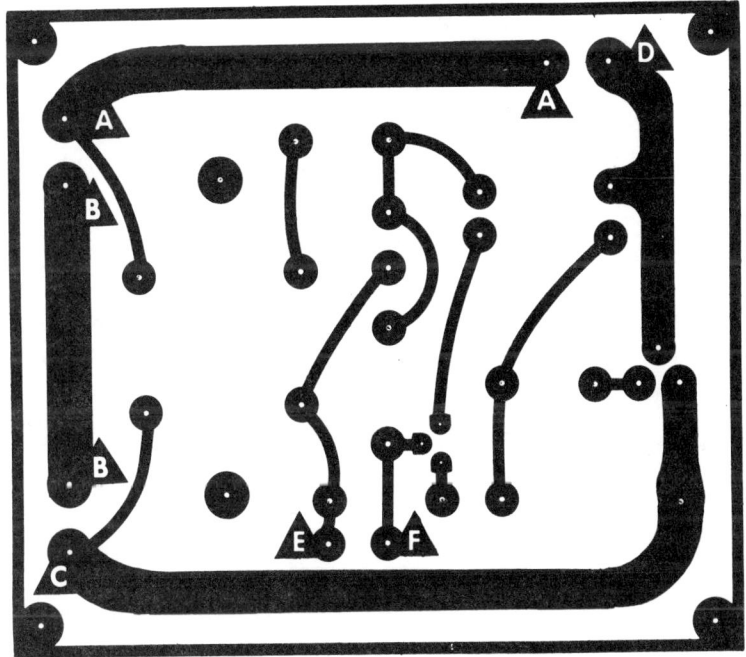

FIGURE 3—4B: TEMPERATURE ACTUATED VENTILA-
TION SYSTEM PC LAYOUT (ACTUAL
SIZE)

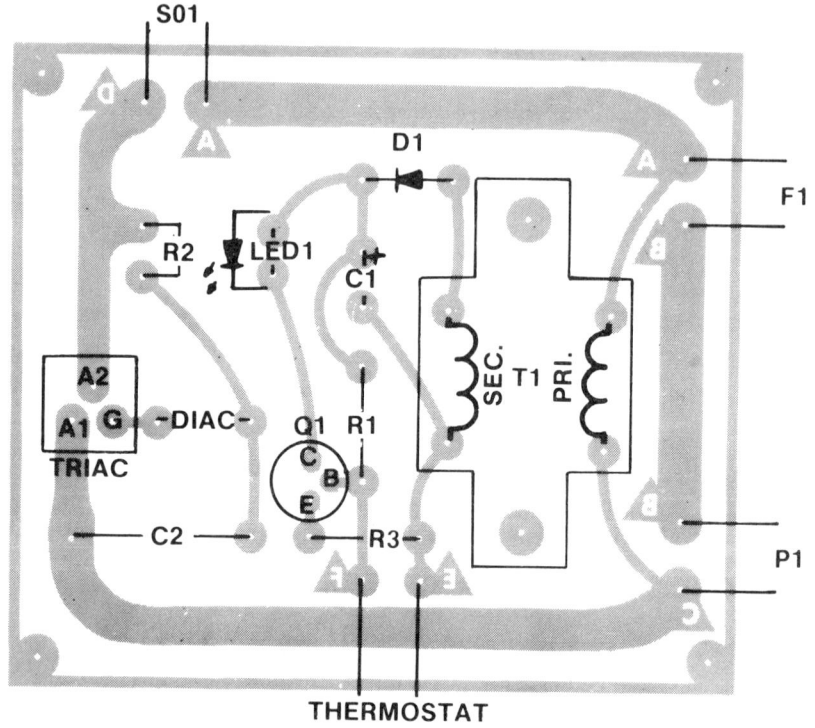

**FIGURE 3—4C: TEMPERATURE ACTUATED VENTILA-
TION SYSTEM COMPONENT PLACE-
MENT GUIDE (TOP VIEW)**

Helpful Construction Hints

Carefully insert and solder the component parts in the printed circuit board (see Figure 3—4C), noting the lead placement of the diode, LED, transistor, and triac. (Refer to Figure 3—4A.) Since LED1 and R2 form an opto-isolator, position the LED to allow maximum light transfer to the photocell's surface. To utilize the full power handling capacity of the triac, a heat sink should be attached to its mounting tab. This tab is electrically isolated and may therefore be directly attached to the enclosure. To increase thermal conductivity, application of silicon grease is recommended.

Any two-terminal replacement thermostat can be used. The thermostat may be mounted on the enclosure or remotely connected with two-conductor bell wire. Since ambient light will affect the photocell, be certain that the minibox is light-proof. This is a line operated device, so be sure that all circuitry is electrically isolated from the enclosure.

How It Works

T1 reduces and isolates the line voltage applied to the half-wave rectifier circuit consisting of D1 and C1. D1 and C1 respectively rectify and filter the voltage required to bias Q1. With the thermostat contacts open (above the selected temperature), R1 provides a positive potential on the base of Q1. This positive voltage forward biases the transistor into conduction, illuminating LED1. Light striking the photocell's surface from LED1 causes the resistance of R2 to decrease. The low resistance of R2 now permits C1 to reach the breakdown voltage of the diac, which triggers the triac into conduction. The triac now supplies power to the series connected outlet SO1. When the preselected temperature has been reached, the thermostat contacts will close, and place a negative potential on the base of Q1. This negative potential reverse biases the transistor and subsequently removes power to the LED. In the absence of light, the photocell's resistance will increase and prevent the capacitor from reaching the diac's breakdown voltage. Thus, power will not be available to the externally connected fan.

Operating Tips

Connect the ventilating fan to SO1 and apply power to the temperature control unit. Adjust the thermostat for the desired setting to actuate the system. When the selected temperature is reached, power will be supplied to the fan. Motors rated up to 400 watts may be controlled by the temperature actuated control unit.

3—5: AUTOMATIC SPRINKLING SYSTEM

What It Does

Tired of the dried out, burned out, lackluster lawn? Hesitant to water your greenery because of exorbitant water costs and the bother of lawn sprinkling? Build and install an automatic sprinkler system and your lawn will become the envy of the neighborhood. Magically, your lawn will receive the precise amount of water it needs, and at the proper time, even when you are away. Let electricity save you time and money—automatically.

Here Are the Schematics

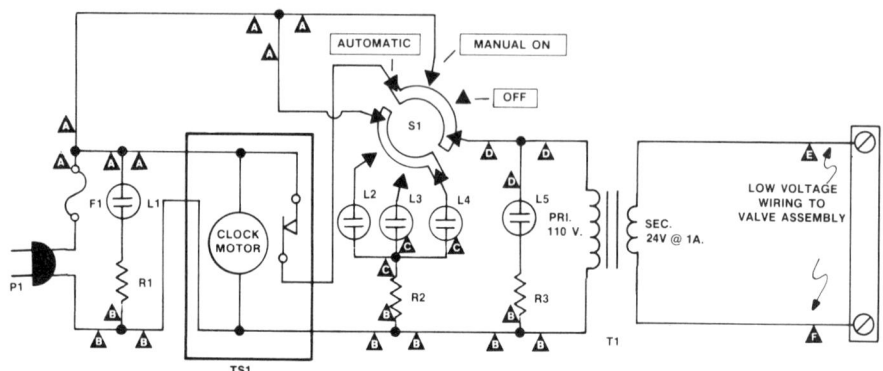

**FIGURE 3—5A: AUTOMATIC   SPRINKLING   TIMER   AS-
SEMBLY**

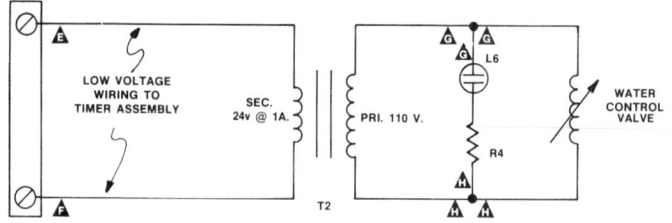

**FIGURE 3—5B: AUTOMATIC   SPRINKLING   CONTROL
VALVE ASSEMBLY**

Parts to Use

F1 – 2 a. fuse	R1-R4 – 56 K ohm, ½ watt,
L1-L6 – NE-2 neon lights	10%
P1 – 110 VAC plug	S1 – 2 pole, 3 position
	rotary switch

T1.T2 – 24 volt transformer   TS1 – 24 hour time switch
@ 1 a.   (Intermatic D811-B)
Misc. – bell wire, enclosures, fuse holder, flange mount
water faucet, knob, lenses, line cord, mounting
hardware, solenoid control valve (Eaton BK-
25840), terminal strips, various plumbing fittings,
etc.

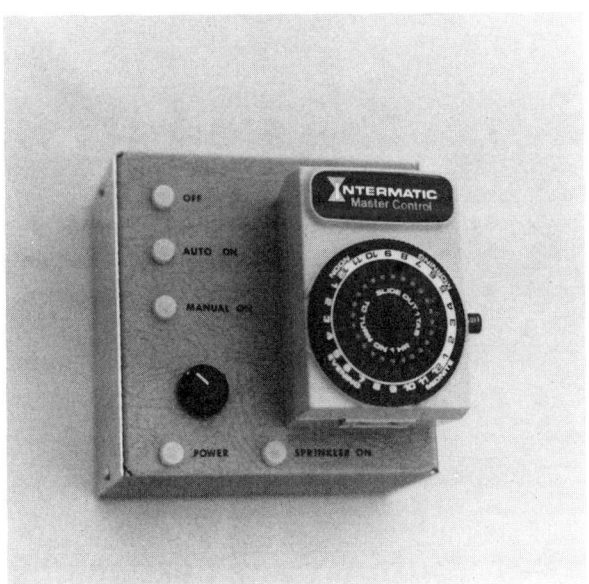

**FIGURE 3—5C: AUTOMATIC SPRINKLING TIMER
ASSEMBLY (FRONT PANEL VIEW)**

Helpful Construction Hints

Point-to-point wiring is suggested, with terminal strips
provided for the display lamps and their respective current limiting
resistors. Be sure that all wiring is electrically isolated from the
enclosure to ensure safe and reliable operation. When securing the
timer to the minibox, avoid interference with the clock mechanism
by carefully locating the mounting screws and internal timer
connections. (See Figure 3—5C.) Interconnect the solenoid valve
and the enclosure faucet with a short length of neoprene tubing

secured with hose clamps. (See Figure 3—5D.) The fitting and adapters required will vary with the style and manufacture of the control valve and faucet used. The control valve is a water inlet valve commonly found in automatic dishwashers, and is available from many appliance parts distributors. A valve with the largest output port should be selected to provide maximum water flow.

FIGURE 3—5D: AUTOMATIC SPRINKLING CONTROL VALVE ASSEMBLY (INTERIOR VIEW)

How It Works (Refer to Figures 3—5A and 3—5B)

At preselected intervals, the timer contacts close, providing power to the primary of T1. The reduced voltage output of T1 is applied to the secondary of T2 and returned to line potential at T2's primary. Receiving line voltage, the solenoid is now energized, opening the water valve. Illumination of L1 indicates power is available to the timer assembly, and lighting of L5 and L6 signifies sprinkler operation. L2, L3, and L4 display the operating mode as selected by S1. T1 and T2 provide safe low voltage operation and

**FIGURE 3—5E: AUTOMATIC SPRINKLING CONTROL
VALVE ASSEMBLY INSTALLATION**

electrical isolation of the remotely connected control valve assembly.

Operating Tips

INSTALLATION: Secure the control valve enclosure near a convenient water outlet. Supply water to the control valve through a short length of garden hose with threaded female couplings attached to each end. (See Figure 3—5E.) When mounting the indoor timer enclosure, allow easy access for timer programming and viewing of the display lamps. Complete the installation by interconnecting the timer and valve assemblies with two-conductor bell wire and providing line voltage to the timer.

OPERATION: Program the timer for the desired length of sprinkler operation and position S1 in the automatic mode. If the recommended timer is incorporated, multiple sprinkling cycles, as short as one hour, can be attained within a 24 hour period. Select the "on" position for S1 if continuous sprinkling is desired, and "off" to disable the system.

3—6: THERMOSTAT CONTROLLER

What It Does

Would you like to reduce your home heating costs easily and inexpensively? Build and install the thermostat controller. Incorporating a low-cost appliance timer and auxiliary thermostat, the controller automatically lowers your home's temperature while you work or sleep. Programmed to return your home to a comfortable temperature before you arrive or awaken, you can save money without discomfort.

Here's the Schematic

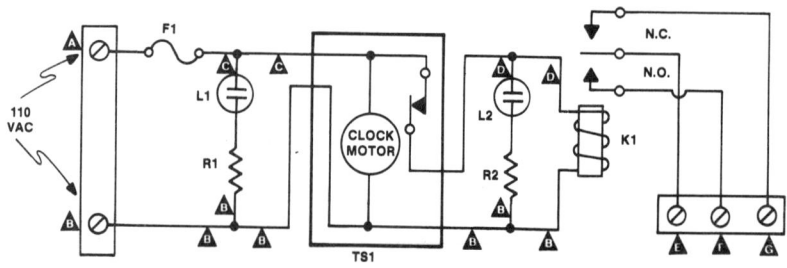

FIGURE 3—6A: TIMER CONTROL UNIT

Parts to Use

F1 – 1 a. fuse	R1, R2 – 56 K ohm, ½ watt, 10%
K1 – SPDT relay 110 VAC coil	Thermostat – see text
L1, L2 – NE-2 neon lights	TS1 – 24 hour timer switch (Intermatic D811-B)
Misc.- barrier terminal strip, bell wire, enclosure, fuse holder, hardware, hookup wire, lamp lenses, etc.	

Helpful Construction Hints

Mount the fuse holder, relay, lamp lenses, and timer in a small enclosure. Point-to-point wiring is suggested, with terminal strips provided for the indicator lamps. To expedite later installation, fasten an electrical workbox cover plate to the minibox and furnish a barrier terminal block for the line wiring. All wires passing through the cover plate should be protected with grommets. Since this is a line operated device, all wiring must be electrically isolated from the enclosure.

How It Works

The timer control unit selects one of two thermostats at preprogrammed intervals. When the timer is programmed "on," K1 is energized as indicated by the lighting of L2. (See Figure 3—6A.) The normally open contacts of K1 now close, switching thermostat B into the gas valve circuit. (See Figure 3—6B.) The temperature setting of this thermostat now controls furnace operation. When the clock timer reaches the "off" mode, power is removed from the relay coil, allowing thermostat A to regulate the heating system. Illumination of L1 signifies that power is available to the timer control unit.

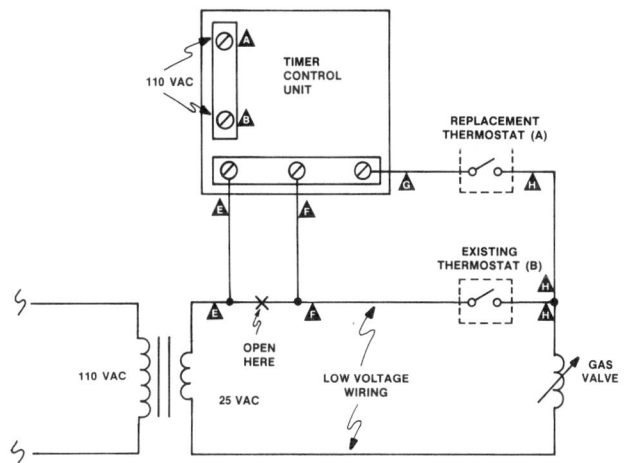

FIGURE 3—6B: TIMER CONTROL UNIT CONNECTED TO EXISTING THERMOSTAT CIRCUIT

Operating Tips

INSTALLATION: Refer to Figure 3—6B. Mount thermostat A (replacement thermostat) next to the existing thermostat. For reliable and accurate operation, be sure to follow the manufacturer's installation instructions explicitly. Select a convenient location for the timer control unit, allowing easy access for timer programming and viewing of the indicator lights.

Next, provide line and low voltage control wiring to an electrical workbox. When installing the 110 VAC supply line, compliance with the local electrical code is recommended. Small gauge three-conductor bell wire is suitable for the control wiring. Complete the installation by connecting the timer unit into the circuit and securing the assembly to the workbox.

OPERATION: Program the timer "on" for a reduced temperature setting, and "off" for normal operation. Depending upon the recovery rate of the heating system, sufficient lead time should be considered in determining the "off" cycle. If increased heating is required during the timer "on" cycle, simply increase the temperature setting of thermostat B. Setting thermostat B for 55 degrees and thermostat A for 68 degrees will substantially reduce heating costs.

4

Testers for
Checking Solid
State Components

4—1: DIODE-RECTIFIER-TRANSISTOR CHECKER

What It Does

You'll find this checker inexpensive, easy-to-build, and just the ticket for checking transistors, rectifiers and diodes out of the circuit. Signal and power transistors can be checked for gain, leakage, shorts and opens. In addition, NPN and PNP types can be determined as well as lead identification.

Diodes' and rectifiers' cathode and anode leads can be determined with just a flick of a switch. It's just as easy to check their forward and reverse resistances. A bonus feature allows the checker to operate as a low voltage continuity tester.

Don't worry about destroying transistors if you happen to connect the leads wrong or have the polarity switch on the opposite type. The transistor will not be harmed, but simply will not check as it should, giving you an indication that it is probably not hooked up correctly.

Here's the Schematic

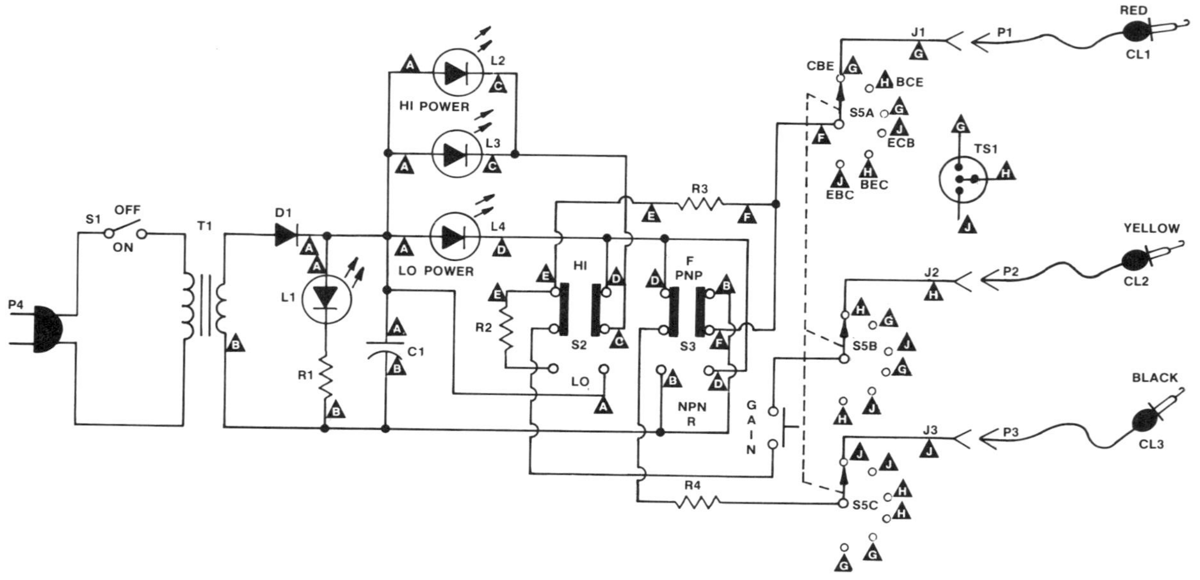

FIGURE 4—1A: DIODE-RECTIFIER-TRANSISTOR CHECKER SCHEMATIC

Parts to Use

C1 – 1000 MFD, 50V	P3 – banana plug, black
CL1 – IC test clip, red	P4 – AC plug
CL2 – IC test clip, yellow	R1 – 270 ohm, $\frac{1}{2}$W, ±10%
CL3 – IC test clip, black	R2, R4 – 100 K$\frac{1}{2}$W, ±10%
D1 – silicon rectifier, 1A	R3 – 2.2K $\frac{1}{2}$W, ±10%
@ 50 PIV	S1 – SPST slide switch
J1 – banana jack, red	S2, S3 – DPDT slide switch
J2 – banana jack, yellow	S4 – SPST push button
J3 – banana jack, black	spring loaded - NO
L1 – LED, green	S5 – 3 poles, 6 position
L2, L3, L4 – LED high	rotary switch
brightness, red	T1 – transformer 117:6.3
P1 – banana plug, red	VAC @ 1A
P2 – banana plug, yellow	TS1 – transistor socket

Misc. – AC line cord, dry transfer labels, enclosure, LED
sockets, terminal strips, test lead wire, etc.

Helpful Construction Hints

Mount the parts on the front panel as shown in Figure 4—1B. A PC board is not needed for this project since most of the connections can easily be made on front panel part lugs.

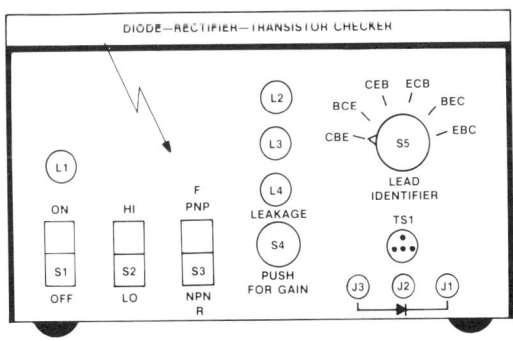

FIGURE 4—1B: DIODE-RECTIFIER-TRANSISTOR
CHECKER FRONT PANEL LAYOUT

Color-code the output jacks and their respective leads and plugs black, yellow and red as shown on the front panel layout. It will make it more convenient when identifying lead placement on a strange transistor. Label the diode-rectifier symbol between the black and red output jack as shown in Figure 4—1B. Recess the high brightness LEDs in the front panel so they can be easily seen when lit in a bright room.

Connecting lines on the lead identifier switch have not been shown on the schematic (see Figure 4—1A) to minimize confusion. Build-O-Matic Guide letters mark wire placement. (Refer to Section 1—1.)

How It Works

The transistor being tested is hooked up in the circuit similar to Figure 4—1C. A low power signal transistor is being tested. When the leakage-gain switch is in the normally open position, the transistor should not conduct. If L4 glows slightly, it is because of emitter-collector leakage. If L4 glows brightly, the emitter-collector junction is shorted. L4 would show no brightness with a good transistor.

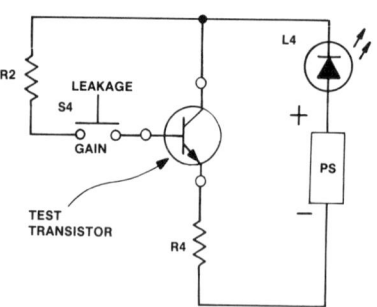

FIGURE 4—1C: SIMPLIFIED CIRCUIT OF TRANSISTOR BEING TESTED

After S4 is pressed, the transistor should conduct causing L4 to glow brightly. Brightness of the LED is a function of transistor gain. If L4 glows dimly, the transistor gain is low. No brightness of L4 indicates a defective transistor.

In the high power position, when checking a high current transistor, all three LEDs (L2, L3, L4) will operate together.

Many high power transistors have considerable leakage so in some cases it is normal for all three LEDs to glow slightly in the leakage position.

Diodes and rectifiers are checked as indicated in the simplified circuit shown in Figure 4—1D. In the reverse mode the diode-rectifier is back biased and the LED should not light. The forward circuit allows the diode-rectifier to conduct causing the LED to glow brightly.

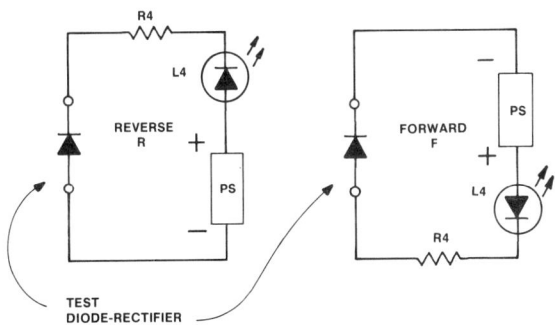

FIGURE 4—1D: SIMPLIFIED CIRCUIT OF DIODE-RECTI-FIER TESTING

Refer to Figure 4—1A for the following circuit explanation. The half-wave power supply is composed of P1, S1, T1, D1, and C1. The ready light (L1) acts as a pilot lamp. The hi-lo switch (S2) is used to change the circuit for high and low power transistors. The PNP-NPN switch (S3) switches power supply polarity for each type of transistor and is used for checking forward and reverse modes of diode-rectifier testing. The lead identifier switch (S5) allows all transistor basing combinations to be tried just by rotating the switch in its six positions. The position that gives the best transistor check is the correct position and also identifies the leads on an unknown transistor. IC test clips and a parallel transistor socket make it convenient to check all types of basing arrangements.

Operating Tips

Never check low power transistors in the HI power position or the transistor being tested may be damaged. Any other combination of switch positions should not harm any device being tested.

When checking a transistor with a known basing arrangement, flick the lead identifier switch to the most clockwise position (EBC). In this position, the black clip will be for the emitter, the yellow clip for the base, and the red clip for the collector. This switch position should also be used when checking diodes and rectifiers. Hook the anode of the diode to the black clip and the cathode to the red clip. (See Figure 4—1B.) If the PNP-NPN switch is in the forward position, a good diode will cause the LO LED to light. Switching to the reverse position will cause the LED to go out.

Here's how to use the lead identifier switch. Suppose you are checking an unknown transistor. As you flick through the various switch positions, try the other controls (NPN-PNP, gain-leakage) to see if you get an indication of a good transistor. If none of the positions works, the transistor is bad. Let's say that the transistor checks good on the third position (CEB) of the lead identifier switch. This means that the transistor lead connected to the black lead is the collector. The yellow lead is clipped to the emitter and the red clip is fastened to the base. Remember that the first letter of the lead identifier switch position always identifies the black clip, the second letter the yellow clip and the third letter the red clip.

To operate the checker as a continuity tester, simply switch the lead identifier switch to the extreme clockwise position. Use the black and red leads. The LO LED will light if there is continuity.

4—2: SCR-TRIAC TESTER

What It Does

Here's a dandy little tester that makes short work of checking SCRs and triacs. It's easy and economical to build and a breeze to use. Just hook up an SCR or triac, flip a few switches and a red-green LED will tell you if it's good or bad.

As an added feature, convenient front panel jacks provide a source of 6.3 VAC and 9 VDC power. The tester can also double as a continuity checker.

Here's the Schematic

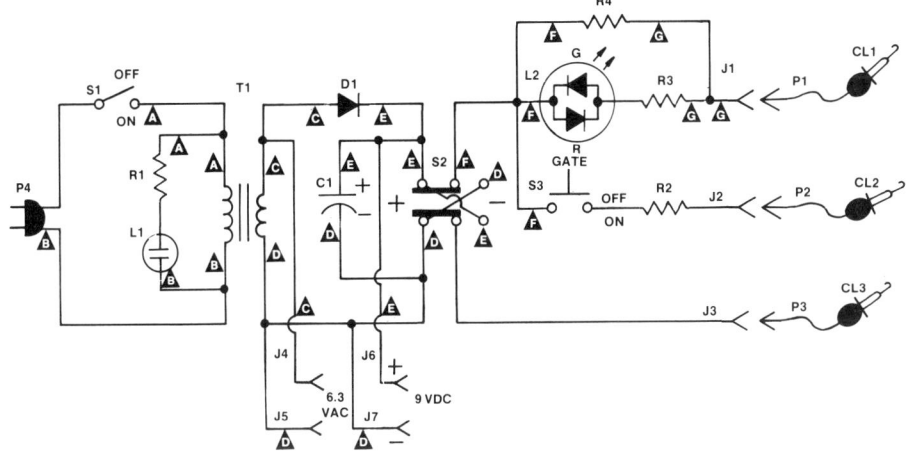

FIGURE 4—2A: SCR-TRIAC TESTER SCHEMATIC

Parts to Use

C1 – 1000 MFD, 50 V	P1 – banana plug, red
CL1 – IC test clip, red	P2 – banana plug, yellow
CL2 – IC test clip, yellow	P3 – banana plug, black
CL3 – IC test clip, black	P4 – AC plug
D1 – silicon rectifier,	R1 – 270 ohm, ½W, ±10%
1A – 50 PIV	R2, R4 – 100 K ½W, ±10%
J1 – banana jack, red	R3 – 2.2K ½W, ±10%
J2 – banana jack, yellow	S1 – SPST slide switch
J3 – banana jack, black	S2 – DPDT slide switch
J4, J5 – binding post,	S3 – SPST momentary
yellow	action push button, NO
J6 – binding post, red	T1 – 117:6.3 transformer
J7 – binding post, black	@ 1A
L1 – NE-2 lamp	
L2 – tri-color LED	

Misc. – enclosure, hardware, LED socket, line cord, neon
    lens and clip, strain relief, terminal strips, etc.

How It Works

Line voltage is stepped down by transformer T1 to 6.3 VAC. (See Figure 4—2A.) It is then rectified (D1) and filtered (C1) to 9 VDC. S1 and L1 act as the on-off switch and pilot light respectively.

Red-green LED (L2) acts as the load for the SCR or triac under test. The LED indicator will glow red when testing a good SCR. When testing a triac the LED will glow red on the first part of the test and green on the second part. Switch (S2) reverses the polarity of the 9 VDC power supply. Gate switch (S3) furnishes a momentary pulse to the gate of the device being tested.

J4 and J5 tap off the 6.3 VAC for availability at the front panel. Likewise J6 and J7 furnish a source of 9 VDC at the front panel.

Helpful Construction Hints

Most of the connections can be made on the part lugs of the front panel components. Use terminal strips for the other connections.

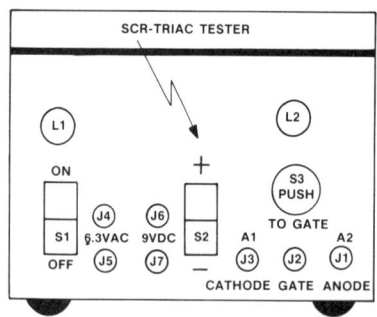

**FIGURE 4—2B: SCR-TRIAC TESTER FRONT PANEL
             LAYOUT**

Figure 4—2B shows the front panel layout for this tester. The project is straightforward to build with no problems. You may want to use a small grommet to hold the red-green LED (L2) instead of an LED socket. Push the LED through the grommet just until the front of the LED pokes out. It makes a nice appearance.

Operating Tips

To check an SCR, hook the black clip to the cathode terminal, the yellow clip to the gate, and the red clip to the anode. Set the polarity reversal switch (S2) to the positive position. Place S1 in the on position. Output indicator LED (L2) should remain out. Momentarily press the gate switch (S3). A good SCR will conduct causing L2 to light red and remain on. Now switch the polarity reversal switch to the negative position. L2 should go out and stay off even if the gate switch is pulsed. Table 4—2A summarizes the SCR checking rules.

S2 POSITION	S3 OFF	S3 ON	S3 OFF
+	L2 OFF	L2 RED ON	L2 RED ON
−	L2 OFF	L2 OFF	L2 OFF

TABLE 4—2A: CHECKING AN SCR

A similar procedure is used to check a triac. This time, connect the black clip to the triac's A1 terminal, the yellow clip to the gate and the red clip to the anode 2 terminal. Set the polarity reversal switch (S2) to the positive position. Place S1 in the on position. Output indicator LED (L2) should remain out. Momentarily press the gate switch (S3). A good triac will conduct causing L2 to light red and remain on. Now switch the polarity reversal switch to the negative position. L2 should go out. Up to this point the checking procedure resembled an SCR test. The triac acts differently from an SCR when the gate switch (S3) is pulsed in the negative position. L2 should light green this time, staying on even when S3 is released. Table 4—2B summarizes the triac checking rules.

S2 POSITION	S3 OFF	S3 ON	S3 OFF
+	L2 OFF	L2 RED ON	L2 RED ON
−	L2 OFF	L2 GREEN ON	L2 GREEN ON

TABLE 4—2B: CHECKING A TRIAC

You might want to duplicate the previous tables and glue them to the top of the SCR-triac tester's enclosure for speedy reference.

Use the red and black clips for checking continuity. If S2 is in the positive position, continuity will be indicated by a red glow from L2. When S2 is in the negative position, L2 will glow green if the continuity is present.

4—3: QUADRAC CHECKER

What It Does

Because of their internal construction, quadracs are difficult to check with conventional testers or by ohmmeter measurements. A quadrac is a combination of a triac and a triggering mechanism called a diac all built into the same package. Refer to the quadrac symbol shown in Figure 4—3A.

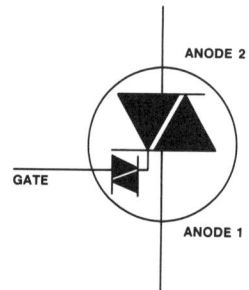

FIGURE 4—3A: QUADRAC SYMBOL

This quadrac checker simulates actual circuit conditions using the quadrac under test to operate the checker. It's easy to test. Simply hook up the quadrac to the checker. Apply power and rotate a potentiometer through its range. A good quadrac will cause an indicator lamp to gradually light from its off condition to full brightness—sweet and simple.

Here's the Schematic

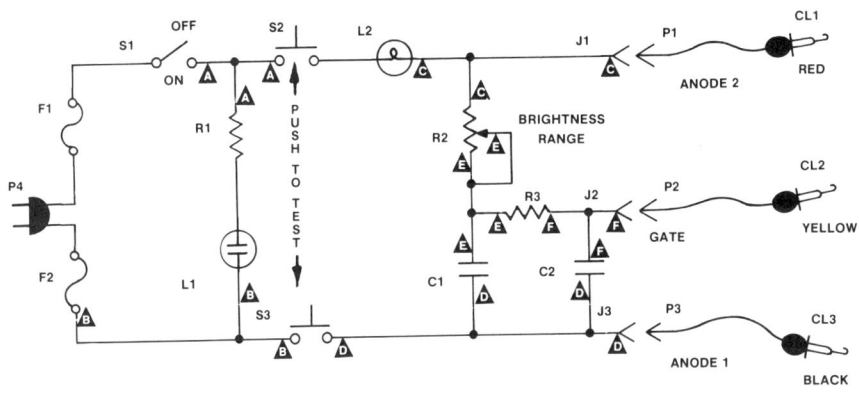

FIGURE 4—3B: QUADRAC CHECKER SCHEMATIC

Parts to Use

C1, C2 – .1 MF, 200V	P1 – banana plug, red
CL1 – IC test clip, red	P2 – banana plug, yellow
CL2 – IC test clip, yellow	P3 – banana plug, black
CL3 – IC test clip, black	P4 – AC plug
F1, F2 – .5A quick blow fuse	R1 – 33K ½W, ±10%
J1 – banana jack, red	R2 – 100K linear poten- tiometer
J2 – banana jack, yellow	R3 – 22K ½W, ±10%
J3 – banana jack, black	S1 – SPST switch
L1 – NE-2 lamp	S2, S3 – SPST momentary
L2 – incandescent lamp 117V, 7W	action push button, NO

Misc. – enclosure, fuseholders, hardware, knob, lamp socket, lens cap and clip, line cord, strain relief, terminal strips, test lead wire, etc.

How It Works

The quadrac being tested controls power to the output indicator L2. (Refer to Figure 4—3B.) Phase shifting network (R2,

C1, C2 and R3) allows the gate clip to control quadrac conduction from off to full on over a uniform range.

F1 and F2, as well as S2 and S3, are operator safety features. In order for the quadrac to be tested, both S2 and S3 have to be pressed simultaneously and the brightness range pot (R2) rotated. This insures that the operator's hands are not touching any part of the quadrac leads that are exposed to line voltage during the test.

Helpful Construction Hints

Mount the front panel components as shown in Figure 4—3C. Most of the connections can be made on the front panel parts' lugs. Use terminal strips for the other connections.

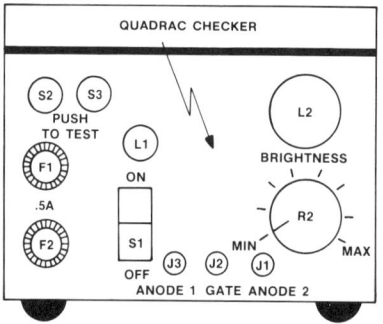

FIGURE 4—3C: QUADRAC CHECKER FRONT PANEL LAYOUT

Be sure to keep push button switches S2 and S3 physically distant from R2 so that two hands are needed to operate the checker. Since line voltage is present on the output jacks, a possible shock will be avoided from touching the leads when testing a quadrac.

A parallel type Christmas tree bulb makes a good indicator lamp. A colored bulb sparkles up the project.

Operating Tips

The quadrac checker is easy to use with no special problems. The only precaution to observe is the gate test clip. Be sure that this clip is connected to the gate and not to one of the other terminals. If it is inadvertently hooked to or touches one of the anodes the quadrac will probably self-destruct.

If a quadrac does not check properly, try interchanging anode 1 and anode 2 test clips. It won't harm anything. Sometimes the anodes are marked improperly or are hard to identify. Figure 4—3D shows a typical quadrac basing arrangement. The heat sink tab is often electrically insulated from all the leads, so don't use it to make one of the anode connections.

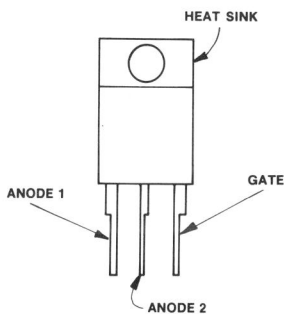

FIGURE 4—3D: TYPICAL QUADRAC BASING ARRANGEMENT

If the checker does not seem to work and you suspect a burned out indicator lamp (L2), just hook the red and black test clips together. Apply power. Push S2 and S3. If the bulb is good it will light at full brightness.

4—4: ZENER DIODE ANALYZER

What It Does

The zener diode analyzer will quickly check and indicate breakdown voltage for any zener from 3.3 V to 30 V. It's easy to build and easy to use.

Let's check a zener diode. Clip the zener to the appropriate leads of the analyzer. Flipping a switch will ignite a green LED showing the test is ready to begin. Momentarily push a rocker switch. The ready light will go out and a red test LED will come on. At the same instant the output meter will read zener voltage.

Releasing the rocker switch returns the analyzer to the ready condition.

Here's the Schematic

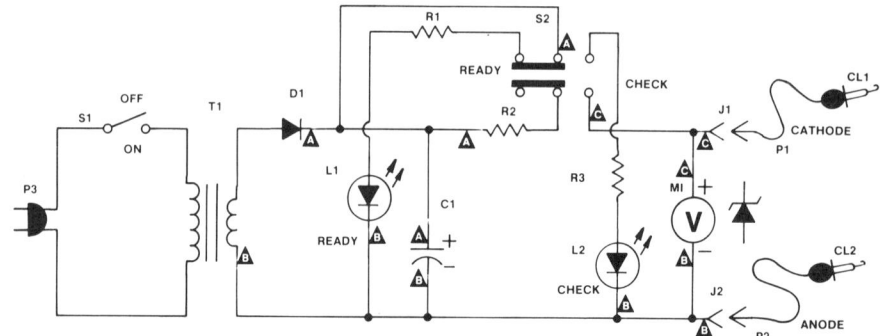

FIGURE 4—4A: ZENER DIODE ANALYZER CIRCUIT

Parts to Use

C1 – 1000 MFD, 50 V	P1 – banana plug, red
CL1 – IC test clip, red	P2 – banana plug, black
CL2 – IC test clip, black	P3 – line cord
D1 – silicon rectifier, 1A @ 100 PIV	R1, R3 – 2.7K 1W, ±10%
	R2 – 470 ohm 5W, ±10%
J1 – banana jack, red	S1 – SPST slide switch
J2 – banana jack, black	S2 - DPDT momentary action rocker slide switch
L1 – LED, green	
L2 – LED, red	T1 – transformer 117:24 @ 1A
M1 – voltmeter, 0-30 VDC	

Misc.- enclosure, LED clips, hardware, line cord, strain relief, test lead wire, terminal strips, etc.

Helpful Construction Hints

Construct the front panel as shown in Figure 4—4B. Most of the connections can be made on the front panel parts' lugs. Use terminal strips for the others. The construction is straightforward with no special problems.

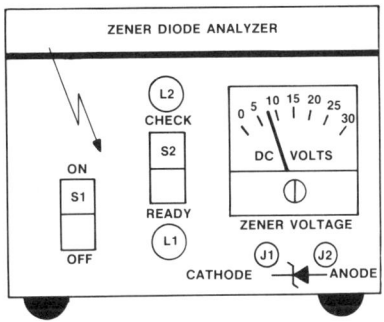

FIGURE 4—4B: ZENER DIODE ANALYZER FRONT PANEL LAYOUT

Be sure to use a momentary action-spring loaded DPDT switch for the ready-check switch (S2). This will insure that no voltage is on the test clips or output meter while the zener is being connected.

It's a good idea to show the zener symbol on the front panel underneath the output jacks for easy reference when hooking up a zener. Use dry transfer labels for a professional appearance.

How It Works

S1 applies power to the step-down transformer (T1) which transforms the 117 VAC line voltage to 24 VAC. Refer to Figure 4—4A. D1 and C1 rectify and filter the transformer's secondary voltage to approximately 34 VDC. The ready LED (L1) will also become activated.

After the test zener is connected to the test clips (CL1 and CL2), the ready-check switch (S2) is pushed. This action will turn off the green ready LED (L1) and activate the red check LED (L2) as well as apply power to the output meter and the zener. The zener will conduct to its zener voltage. The actual zener voltage is then read directly from the output meter. When the ready-check switch is released, it automatically returns to the ready position, lighting the green LED and removing power from the red LED, the meter, and the zener. That's all there is to it—duck soup!

Operating Tips

Be sure to hook the zener's anode and cathode leads to the correct test clips or you'll get misleading meter readings. If you accidentally connect the zener leads backwards, the zener will conduct in the forward direction like an ordinary rectifier. The meter voltage will drop to zero. If this does happen don't be alarmed because neither the zener nor the analyzer will be harmed. Just reverse the leads and try again. In fact, it's a good idea to try checking a zener in both directions if zero meter volts is indicated on the first try. Maybe the zener was marked wrong.

Suppose you are checking a 12 V zener and the meter reads 8 V. You have found a problem but don't just pitch the wrong reading zener. Label it with masking tape and throw it in the junk box. Maybe you'll need an 8 V zener some day.

If the meter reads full scale in one direction and zero in the other, the zener is shot or its zener voltage is higher than 30 V (not likely), or it's not even a zener—a diode would give that kind of indication. A meter reading of zero volts in both directions would indicate a shorted zener. An open zener would give a full scale reading in both directions.

4—5: CURVE TRACER

What It Does

The curve tracer coupled with an oscilloscope can be used to check most types of solid state junctions in or out of the circuit. Figure 4—5A shows oscilloscope patterns that you would expect to see as you check good and bad junctions. For example, suppose you were checking across the emitter-base leads of a silicon transistor. If the waveform resembled Figure 4—5A, #3 or #4, the junction would be okay. However, if it looked like waveforms #1, #2 or #5, the junction would be defective.

Each junction of a solid state device is tested in this manner. It's quick and easy to clip on the test clips while watching the oscilloscope screen. In most cases it's not even necessary to know what junctions you are testing—just look and compare.

FIGURE 4—5A: OSCILLOSCOPE SOLID STATE JUNCTION PATTERNS

Here's the Schematic

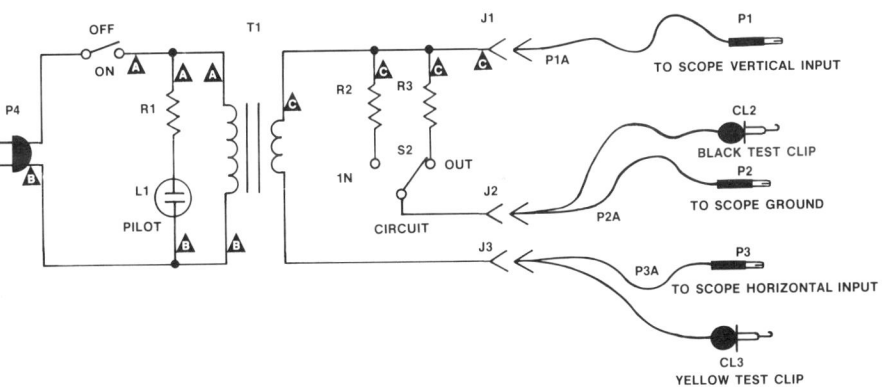

FIGURE 4—5B: CURVE TRACER SCHEMATIC

Parts to Use

CL2 - IC test clip, black
CL3 - IC test clip, yellow
J1 – binding post, red
J2 – binding post, black
J3 – binding post, yellow
L1 - NE-2 neon lamp
P1 – banana plug, red
P2, P2A – banana plug, black

P3, P3A – banana plug, yellow
P4 - AC plug
R1 - 33K ½W, ±10%
R2 - 330 ohm ½W, ±10%
R3 - 3.3K ½W, ±10%
S1 - SPST switch
S2 - SPDT switch
T1 – transformer 117:6.3 VAC

Misc.- enclosure, hardware, lens cap and clip, line cord, strain relief, terminal strips, test lead wire, etc.

Helpful Construction Hints

Many of the parts are mounted on the front panel. Use their lugs and terminal strips for all part connections. This project is very easy to build since the parts count is so small and the circuit is very basic. You shouldn't have any problems with the construction.

Since there are five test leads associated with this project, don't scrimp on their length. Give the leads that plug into the scope plenty of length so they don't interfere with the other two test clip leads that are used for checking parts' junctions.

For ready reference it's a good idea to display the patterns shown in Figure 4-5A on the front panel of the curve tracer. Copy the patterns on light cardboard and glue in place or make your own patterns with dry transfer symbols. It puts a nice finishing touch on the project.

How It Works

A solid state junction is made where two semiconductor materials are joined. Under certain conditions the junction will act as an insulator (open) and with other conditions it acts as a conductor (short). Reviewing scope patterns shown in Figure 4—5A illustrates that good junction patterns resemble an open and a short connected together to form a right angle. One of the "good" patterns is inverted simply because the curve tracer's test clips were interchanged or the opposite semiconductor material was checked.

A good junction should show a horizontal line during the period it is not conducting and then an abrupt 90° vertical line when it begins conduction. If the change from horizontal to vertical, junction breakdown, is gradual as shown in pattern #5, it represents a leaky junction.

Refer to Figure 4—5B for the following discussion. S1 applies power to the step-down transformer (T1) and to the pilot lamp (L1). 6.3 VAC is applied across a series combination of a resistor and a solid state junction. The vertical input terminals of the scope displays the waveform across the resistor while scope sweep is produced by the voltage dropped across the junction. Switch S2 decreases the resistance of the curve tracer tenfold for better in-circuit measurements.

Operating Tips

To prepare the oscilloscope for curve tracing, connect P1 (see Figure 4—5B) to the scope's vertical input jack. Feed P2 of the curve tracer to the ground terminal of the scope. Plug P3 into the scope's horizontal input jack. Set the scope's horizontal sweep control to external input. Apply power to the scope. While watching the scope screen, short the yellow and black test clips, (CL2, CL3) together. Adjust the scope controls until the pattern resembles #1 in Figure 4—5A. Disconnect CL2 and CL3. Now the waveform should resemble #2 pattern. Readjust scope controls for optimum pattern. You are now ready to begin junction curve tracing.

When testing transistor junctions you will get meaningful waveform patterns for the base-emitter and for the base-collector. However, in many transistors it is very difficult to get a "good" pattern from the emitter-collector junction. Increasing the gain of the vertical scope input will often show evidence of a slight junction breakdown.

Although the best patterns are achieved with out-of-circuit testing, it is also possible to obtain meaningful waveforms in-the-circuit. Be sure that circuit power is off before connecting the curve tracer to a junction. In many cases the oscilloscope pattern will not be exactly like the waveforms shown in Figure 4—5A. This is because there are other parts in parallel with most of the solid state junctions. Each part will influence the pattern depending on its electrical characteristics.

The best and most meaningful way to capitalize on in-the-circuit curve tracing is to prepare ahead of time "signature" patterns for all junctions in the equipment that you routinely troubleshoot. In this way you will know exactly the type of patterns to expect when checking junctions in defective equipment. This is certainly not practical for technicians who service a wide variety of apparatus, but for those who specialize on a particular line of equipment it really pays off in fast troubleshooting.

4—6: LIGHT-EMITTING DIODE TESTER

What It Does

If you have ever tried to test a seven-segment LED display, you will appreciate this handy project. Eliminate the exasperation of struggling with test leads, and appropriate current-limiting resistors. With the built-in test socket, the LED checker can accommodate a wide variety of displays. In addition, a handy set of test clips facilitates testing of LED indicators.

Here's the Schematic

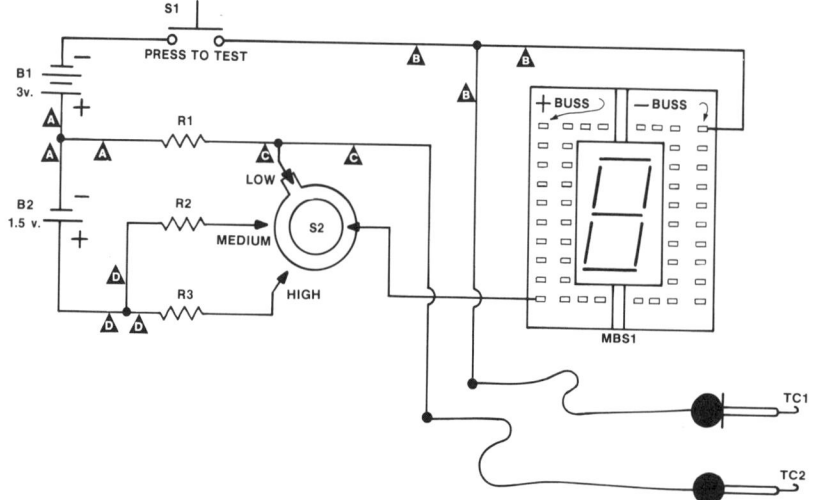

FIGURE 4—6A: LIGHT-EMITTING DIODE TESTER

Parts to Use

B1 – 2 series connected penlight cells	R2 – 33 ohms, ½ watt, 10%
	R3 – 15 ohms, ½ watt, 10%
B2 – single penlight cell	S1 – SPST NO push button switch (spring loaded)
MBS1 – modular bread-board socket (Radio Shack 276-176)	S2 – 1 pole, 3 position rotary switch
R1 – 100 ohms, ½ watt, 10%	

TC1 - black mini test clip   TC2 – red mini test clip
(retractable)   (retractable)
Misc. - battery holder, enclosure, grommets, hardware, hookup wire, knob, etc.

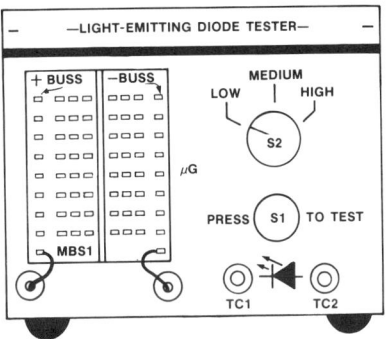

FIGURE 4—6B: LIGHT-EMITTING DIODE TESTER (FRONT PANEL VIEW)

Helpful Construction Hints (See Figure 4—6B)

Mount the switches, battery holder, and breadboard socket in a small enclosure. For a more professional appearance, attach the socket to the rear of the front panel, and provide a rectangular opening in the minibox, allowing access to the tie points. The socket may be secured with machine screws and nuts or epoxy cement. If epoxy is used, tightly clamp the breadboard to the panel and allow at least 24 hours drying time. Connection to the socket's buss lines and test clips are made through openings in the front panel. To insure reliable operation, protect these leads with small grommets.

How It Works (Refer to Figure 4—6A)

Closing S1 applies power to the breadboard socket and the test clips. This socket allows convenient testing of seven-segment displays through its tie-point busses. Since the current requirements of such displays differ, S2 provides selectable current limiting through R1, R2, or R3. LEDs connected to the test clips received their power from B1, with current control provided by R1.

Seven-segment display (size in inches)	S2 position
.125	low
.25 – .4	medium
.5 – .8	high

FIGURE 4—6C: S2 POSITION CHART

Operating Tips

LEDs: Connect the cathode and anode leads to TC1 and TC2 respectively. Depress S1 to test the LED.

SEVEN-SEGMENT DISPLAYS: Insert the display into MBS1. *Common Cathode* – connect the common leads to the negative buss with short lengths of 24 gauge solid wire. Wire each of the remaining segment terminals to the positive buss line. *Common Anode* – Common leads to positive buss, remaining terminals to the negative buss.

Select the proper position for S2 (see Figure 4—6C) and press S1 to test the display. Check for even illumination of all display segments.

4—7: FIELD EFFECT TRANSISTOR CHECKER

What It Does

Have you ever bypassed a bargain priced grab bag assortment of electronic components because you had no easy method for testing them? By adding this invaluable field effect transistor checker to your test equipment, you can now take fuller advantage of the plentiful supply of surplus components. In addition to the money-saving benefits of the FET checker, it will prove to be a great timesaver in troubleshooting. The condition of junction as well as most MOSFETs can be effortlessly determined. The FET checker quickly and easily separates the good parts from the bad, and it is constructed of readily available and inexpensive parts.

Here's the Schematic

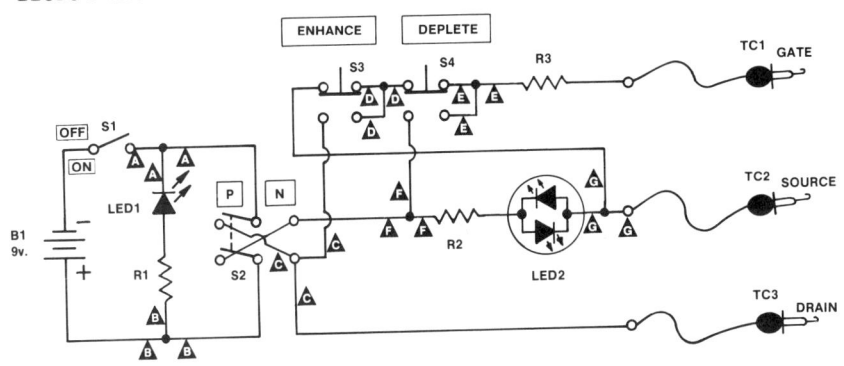

FIGURE 4—7A: FIELD EFFECT TRANSISTOR CHECKER

Parts to Use

B1 – 9 volt transistor radio battery
LED1 – jumbo red
LED2 – tri-color (Radio Shack 276-035)
R1, R2 – 270 ohm, ½ watt, 10%
R3, – 1 K ohm, ½ watt, 10%
S1 – SPST switch
S2 – DPDT switch
S3, S4 – DPST push button switch (spring loaded)
TC1, TC2, TC3 – mini test clips (retractable)
Misc. - battery connector, enclosure, grommets, hard-
ware, hookup wire, terminal strips, test lead wire,
etc.

Helpful Construction Hints

Install the switches, LEDs, etc. in a small enclosure. Point-to-point wiring is suggested, with terminal strips provided for the LEDs and test clip leads. Protect the leads passing through the front panel with small grommets and knot each wire to provide strain relief. Be sure that the long lead of LED2 is connected to guide letter G. If properly wired, the LED will glow red and green, respectively, when checking N and P channel FETs. Spring loaded switches are

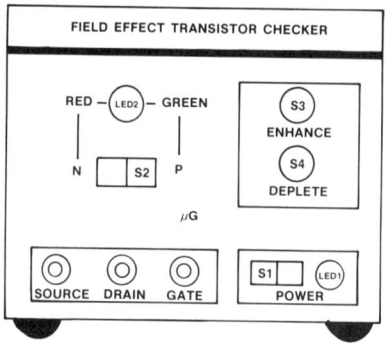

FIGURE 4—7B: FIELD EFFECT TRANSISTOR CHECKER (FRONT PANEL VIEW)

required for S3 and S4, with their normally closed contacts wire to guide letters D, E, and G.

How It Works (See Figure 4—7A)

S1 applies checker power, with LED1 serving as a "power on" indicator. S2 selects both the proper biasing applied to the source-drain circuit, and the correct polarity to the normally open contacts of S3 and S4. When S3 and S4 are normally closed, the source and gate leads are maintained at the same electrical potential, causing a depletion FET under test to exhibit a low source-drain resistance. This low resistance allows sufficient current to illuminate LED2. R2 maintains a safe current level in the event the source and drain are shorted. Depressing S4 pinches off the source-drain. This increased channel resistance causes LED2 to dim or extinguish. Enhancement FETs will exhibit a high source-drain resistance until the gate is removed from the source potential by depressing S3. R3 limits the current applied to the gate of junction FETs when S4 is pressed.

Operating Tips (Refer to Figure 4—7C)

Connect the test clips to the appropriate FET leads. *Note*: To prevent possible transient voltage damage to the field effect transistor, checker power should remain off when attaching or disconnecting the test clips. Closing S1 will apply circuit power as indicated by illumination of LED1. Depletion FETs will cause

FET Type	S2	S3	S4	Remarks
N-channel (junction)	N		press to test	LED2 will glow red; extinguish when S4 is depressed.*
P-channel (junction)	P		press to test	LED2 will glow green; extinguish when S4 is depressed.*
N-channel (depletion-MOS)	N		press to test	Same as N-channel junction FET
P-channel (depletion-MOS)	P		press to test	Same as P-channel junction FET
N-channel (enhancement-MOS)	N	press to test		LED2 off; glows red when S3 is depressed.
P-channel (enhancement-MOS)	P	press to test		LED2 off; glows green when S3 is depressed.
				*NOTE: Some FETs will cause LED2 to only decrease in brilliance.

FIGURE 4—7C: FIELD EFFECT TRANSISTOR CHECKER OPERATION CHART

LED2 to light either red or green, depending upon the channel type. Depressing S4 will dim or extinguish the LED. A good enhancement FET will not illuminate LED2 until S3 is pressed. Because of the susceptibility of MOSFETs to static charges, special care is necessary in their testing and handling. Testing should be conducted on a grounded metal plate, with the operator's arms contacting the plate's surface. Immediately after testing, be sure to reinstall the FET's protecting foil or shorting ring.

4—8: DIAC CHECKER

What It Does

The diac, a triggering mechanism for a triac, is very difficult to check with conventional test instruments. It is a bidirectional switching device that offers a high circuit impedance until a certain voltage level is reached. It will then break down, significantly decreasing its impedance, allowing a fair amount of current to flow.

This checker uses an actual operating circuit to check diacs. The diac being checked is temporarily hooked into the gate circuit of a triac-operated lamp dimmer. A good diac will enable the brightness of the bulb to vary uniformly from off to full on. Operation of the checker is a snap. Just hook up the diac. Push a couple of push buttons; turn a potentiometer while observing bulb brightness range. That's all there is to it.

Here's the Schematic

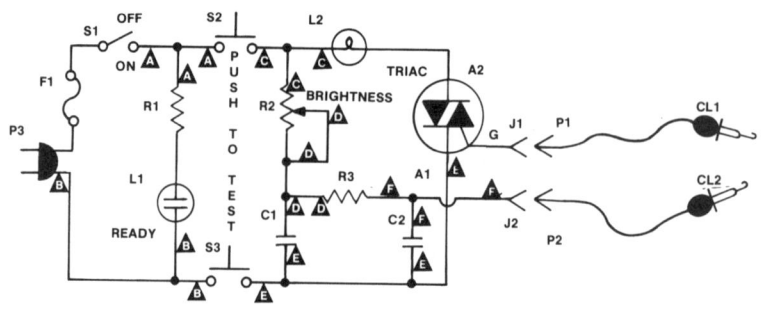

FIGURE 4—8A: DIAC CHECKER SCHEMATIC

Parts to Use

C1, C2 – .1 MF, 200 V	P3 – AC plug
CL1, CL2 – IC test clip, yellow	R1 – 33 K ½W, ±10%
	R2 – 100k linear potentiometer
F1 – .5A fast blow fuse	R3 – 22K ½W, ±10%
J1, J2 – banana jack yellow	S1 - SPST switch
	S2, S3 - SPST momentary push button switch, NO
L1 – NE-2 neon lamp	
L2 – lamp, 117V - 7W	TRIAC – triac, 2A @ 200 PIV
P1, P2 – banana plug, yellow	

Misc. – enclosure, fuseholder, lamp socket, lens cap and clip, line cord, strain relief, terminal strips, test lead wire, etc.

Helpful Construction Hints

Mount the parts on the front panel similar to Figure 4—8B. A printed circuit is not needed since most of the connections can be made on the front panel parts' lugs. Use terminal strips for the others.

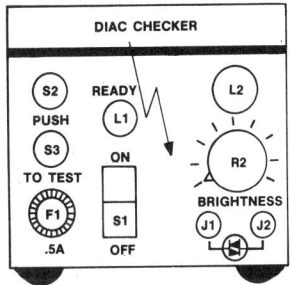

FIGURE 4—8B: DIAC CHECKER FRONT PANEL LAYOUT

Construction is easy with no special problems. A parallel type Christmas tree bulb works well for the incandescent bulb. If you don't want to bother with a lamp socket, simply use a large rubber grommet. Poke the bulb 1/3 of the way through the grommet, letting the colored bulb act as the lens. Several dabs of silicon rubber on the inside of the chassis help to hold the bulb firmly in place. Make the bulb connections directly to its base terminals.

How It Works

Closing S1 applies power to the pilot lamp (L1) and to the rest of the lamp dimmer circuit through push button switches S2 and S3. (Refer to Figure 4—8A.) S2 and S3 keep line voltage potential off of the output clips except during test. R2, C1, R3, and C2 act as a phase-shifting network for the gate circuit of the triac. L2 and the triac are in series across the AC line, sharing line voltage depending on the amount of triac conduction. The setting of the brightness pot (R2) determines when the test diac will fire in turn triggering the triac.

When the diac fires early in the AC cycle, the bulb (L2) will show near full brightness. Firing later in the cycle causes the bulb brightness to decrease. The brightness becomes very dim when the diac fires late in the cycle.

Operating Tips

Hook the test diac with either polarity to the diac checker. Since a diac is a bidirectional switching device, polarity is not a factor.

Press the two push button switches with one hand while rotating the potentiometer with the other. This will ensure that your hands will not come in contact with line potential during the diac test.

If the bulb does not turn on uniformly in brightness but rather comes on in steps, you probably have a shorted diac. An open diac will cause zero bulb brightness. A good diac should cause the brightness of the bulb to light uniformly from full off to full on.

4—9: UNIJUNCTION TRANSISTOR TESTER

What It Does

It talks to you! That's right. A good unijunction transistor will produce a unique buzz when it is checked. A defective one will be very quiet.

Just hook up the two base leads and the emitter to the UJT checker and apply power. Rotate a potentiometer back and forth to hear the characteristic buzz of a good unijunction transistor. It's that simple. No fuss, no muss.

Here's the Schematic

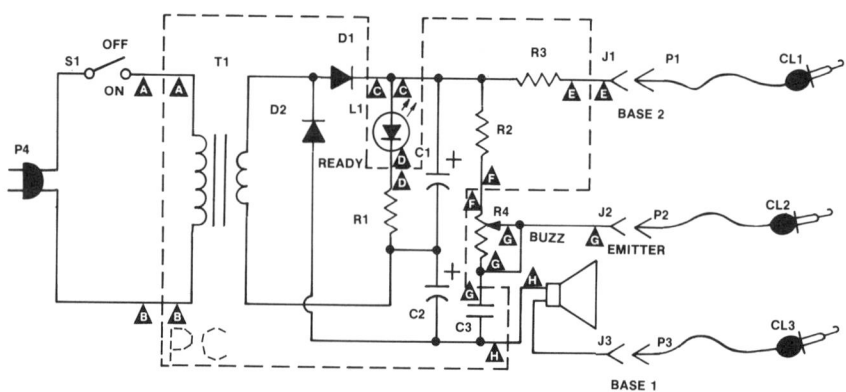

FIGURE 4—9A: UNIJUNCTION TRANSISTOR TESTER SCHEMATIC

Parts to Use

C1, C2 – 1000 MFD, 50 V	P1 – banana plug, red
C3 – .5 MF, 50 V	P2 – banana plug, yellow
CL1 – IC test clip, red	P3 – banana plug, black
CL2 – IC test clip, yellow	R1 – 270 ohm ½W, ± 10%
CL3 – IC test clip, black	R2 – 33K ½W, ± 10%
D1, D2 – silicon rectifier	R3 – 390 ohm ½W, ± 10%
.5A – 50 PIV	R4 - 10K linear poten-
J1 – banana jack, red	tiometer
J2 – banana jack, yellow	S1 – SPST switch
J3 – banana jack, black	SPK – speaker, 8 ohm – 4"
L1 – LED	T1 – transformer 117:6.3
	VAC

Misc.- enclosure, hardware, knob, LED clips, line cord, strain relief, terminal strips, test lead wire, etc.

The PC Layout You'll Need

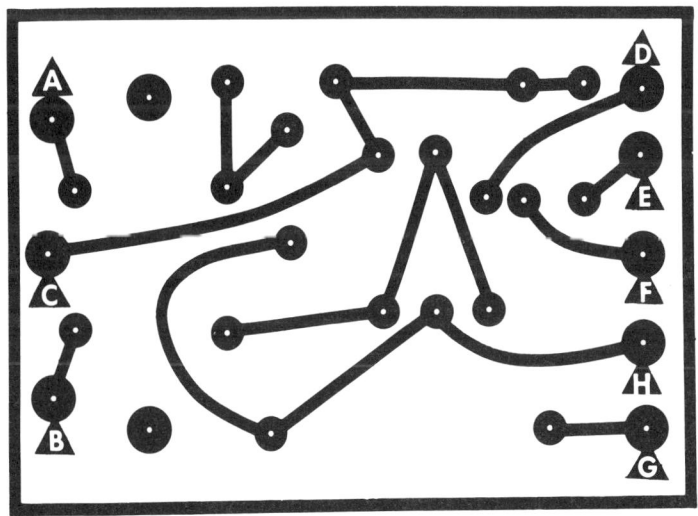

FIGURE 4—9B: UJT TESTER PC LAYOUT (ACTUAL SIZE)

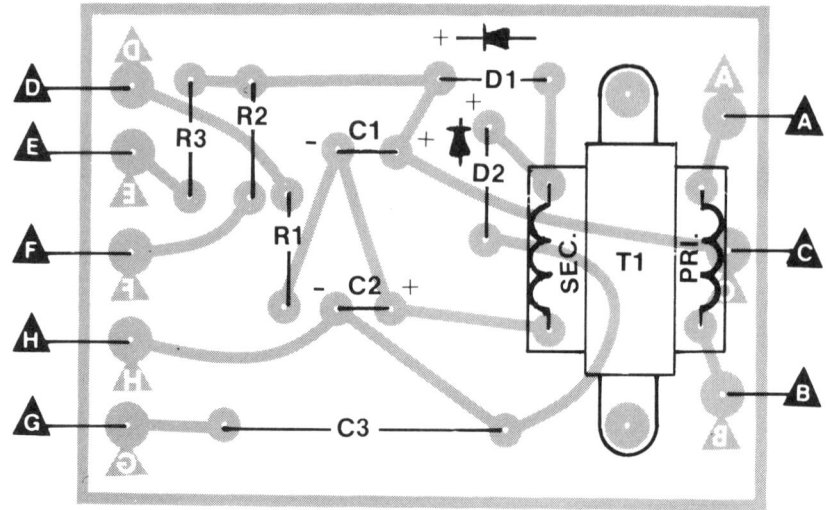

**FIGURE 4—9C: UJT TESTER COMPONENT PLACEMENT
GUIDE (TOP VIEW)**

Helpful Construction Hints

Lay out the front panel as shown in Figure 4—9D. Use the front panel parts' lugs and the PC board's input-output pads for all tie points.

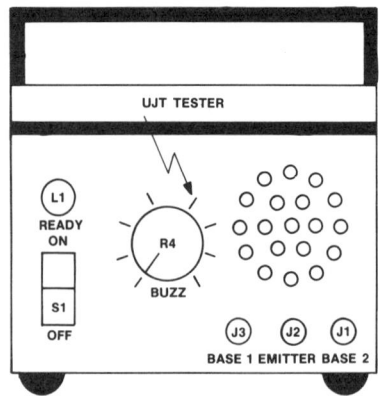

FIGURE 4—9D: UJT TESTER FRONT PANEL LAYOUT

A pattern of holes can be drilled for the speaker grill. Be sure to drill enough holes to allow the UJT buzz to be heard without straining an eardrum. Probably a better method would be to cut one larger speaker hole in the front panel, covering it on the inside with perforated metal, grill cloth or sponge.

How It Works

Refer to Figure 4—9A for the following explanation. S1 applies power to the 117:6.3 VAC step-down transformer. D1 and D2 coupled with C1 and C2 form a voltage doubler circuit that produces approximately 18 VDC for the rest of the tester. The ready LED (L1) acts as a pilot lamp.

The UJT under test becomes the heart of a relaxation oscillator. R2, R4, and C3 form an RC timing network. Potentiometer R4 controls the frequency of the oscillator and the sound of the speaker buzz. A defective UJT disables oscillator action, resulting in no sound output.

Operating Tips

If base 1 and base 2 are interchanged, the buzz will still be audible. However, the volume and crispness of the sound will not compare to the correct connections. This feature allows easier identification of an unknown UJT basing arrangement. The checker or the UJT will not be harmed with the wrong test clip hookups.

Here's an exampale of random checking a UJT with an unknown basing arrangement:

1. Base 2 correct, emitter and base 1 interchanged—no sound.
2. Base 1 correct, emitter and base interchanged—no sound.
3. Emitter correct, base 1 and base 2 interchanged—weak sound.
4. Emitter, base 1 and base 2 correct—good buzz. The UJT has been checked and the leads identified!

Sometimes the oscillator will operate for only a portion of the buzz control pot's range. For this reason be sure to operate the pot over its entire range for each basing combination.

5

Projects to
Make Life
Easier

5

5—1: MOISTURE-WATER DETECTOR

What It Does

Here's a very sensitive electronic circuit that will automatically detect the presence of moisture or water. As soon as moisture is discovered, a piercing 50 db sound will be emitted and a 117 VAC outlet will become energized.

The moisture-water detector is an ideal device to monitor a basement floor that is subject to flooding. Simply plug a sump pump into the detector's outlet. At the slightest trace of floor water the detector will become activated. The sump pump motor will start pumping water out while a shrill audible signal will notify everyone in the house of the problem.

A swimming pool splash alarm, water level detector, rain alert, overflow indicator are just a few of the many applications that the moisture-water detector could be used for.

Here's the Schematic

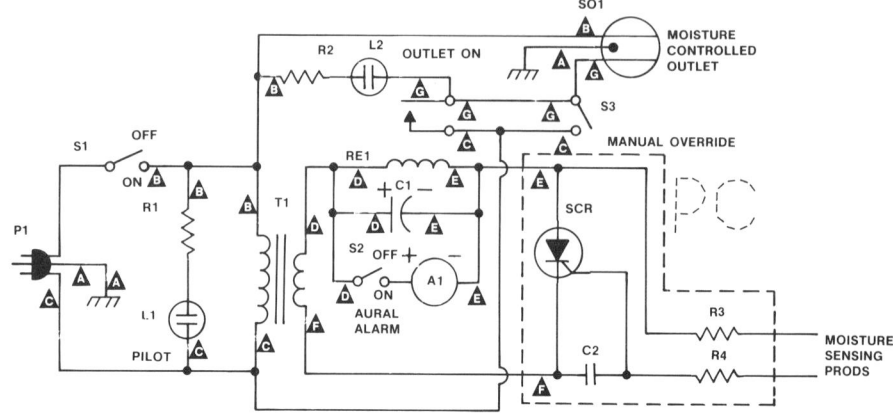

FIGURE 5—1A: MOISTURE-WATER DETECTOR SCHEMATIC

Parts to Use

A1 – piezoelectronic transducer, 24 VDC

C1 – 10 MF, 50 V electrolytic

C2 - .01 MF, 50 V

L1, L2 - NE-2 lamp

P1 - AC plug with safety ground

R1, R2 - 33K ½W, ±10%
R3, R4 – 18K ½W, ±10%

RE1 – SPST relay, 24 VDC coil with 10A contacts

S1, S3 - SPST switch, 117 VAC with 10A contacts

S2 - SPST switch, 24 VDC with 1A contacts

SCR - C103Y SCR, 800 MA @ 60 V

SO1 - AC outlet with safety ground

T1 – transformer, 117:24 VAC

Misc. - AC 3 conductor cord, hardware, mounting clip, neon lens caps, pill vial, probe cable—2 conductor stranded, silicon rubber, strain reliefs, terminal strips, etc.

The PC Layout You'll Need

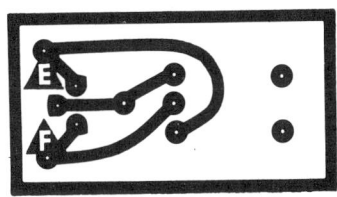

FIGURE 5—1B: MOISTURE-WATER DETECTOR PROBE PC LAYOUT (ACTUAL SIZE)

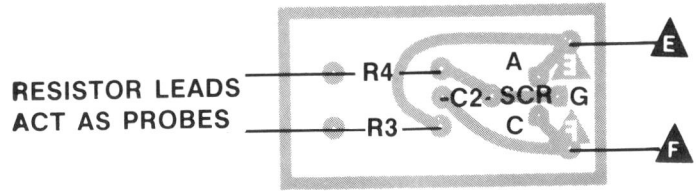

FIGURE 5—1C: MOISTURE-WATER DETECTOR PROBE COMPONENT PLACEMENT GUIDE (TOP VIEW)

Helpful Construction Hints

Figure 5—1D shows the front panel layout of the moisture-water detector. It is designed to be wall mounted close to the monitored area. PC construction is not suitable for the main part of this project since most of the parts used are mounted on the chassis and rather bulky. The inside of the detector is shown in Figure 5—1E. Use three terminal strips; one each for the neon lamp assemblies and the other for the transformer and line cord connections. The remainder of the connections can be made on the part lugs.

The probe assembly is the most tricky part of construction. Use a small pill vial for the probe case. Cut the probe PC board (see Figure 5—1B) so that it just slides into the pill vial. Drill two 1/32" holes through the bottom of the vial so the ends of R1 and R2 will fit through as the PC is pushed into the vial. The resistor leads act as the sensing prods.

After the probe assembly (Figure 5—1F) has been completed, it must be made completely waterproof. Silicon rubber liberally dabbed around all openings does a good job. Just to make sure you

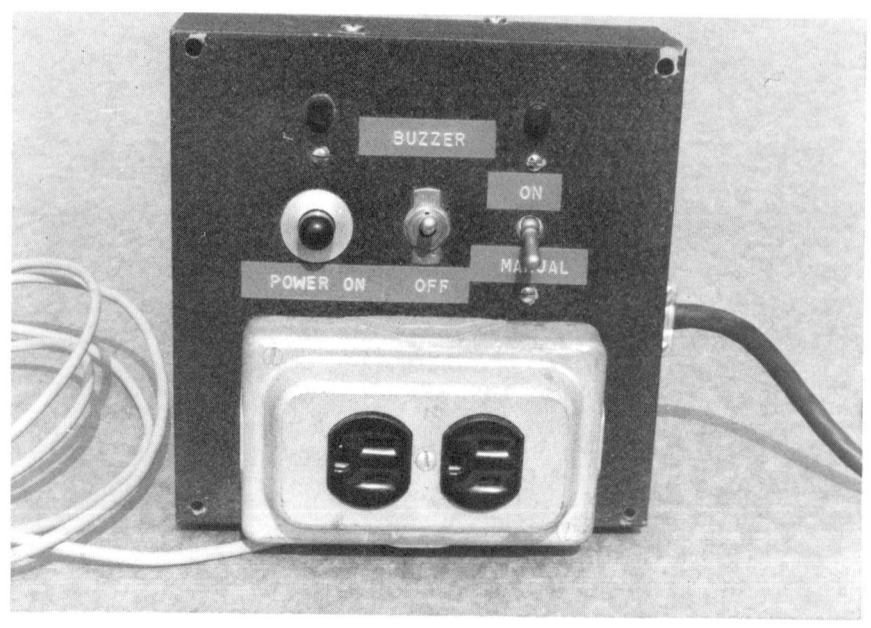

**FIGURE 5—1D: MOISTURE-WATER DETECTOR
(OUTSIDE VIEW)**

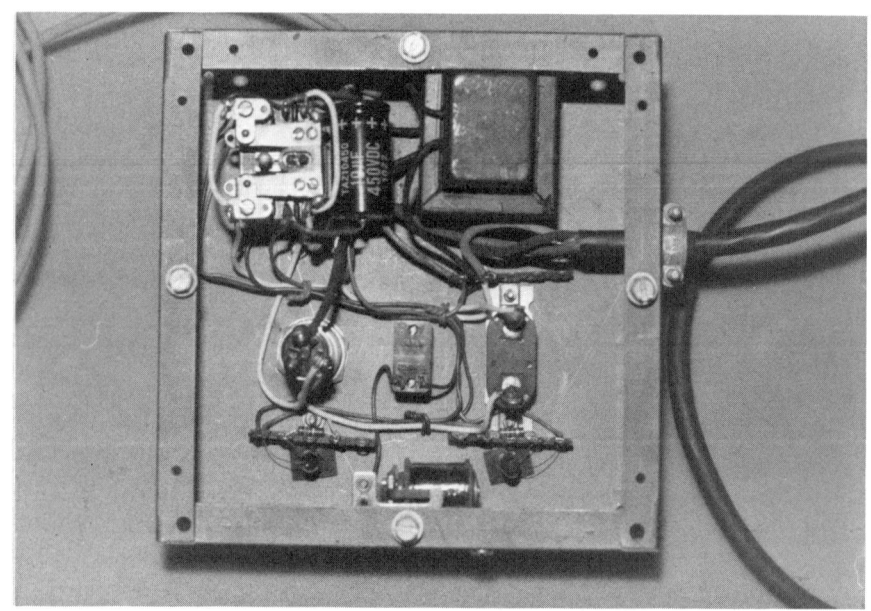

**FIGURE 5—1E: MOISTURE-WATER DETECTOR (INSIDE
VIEW)**

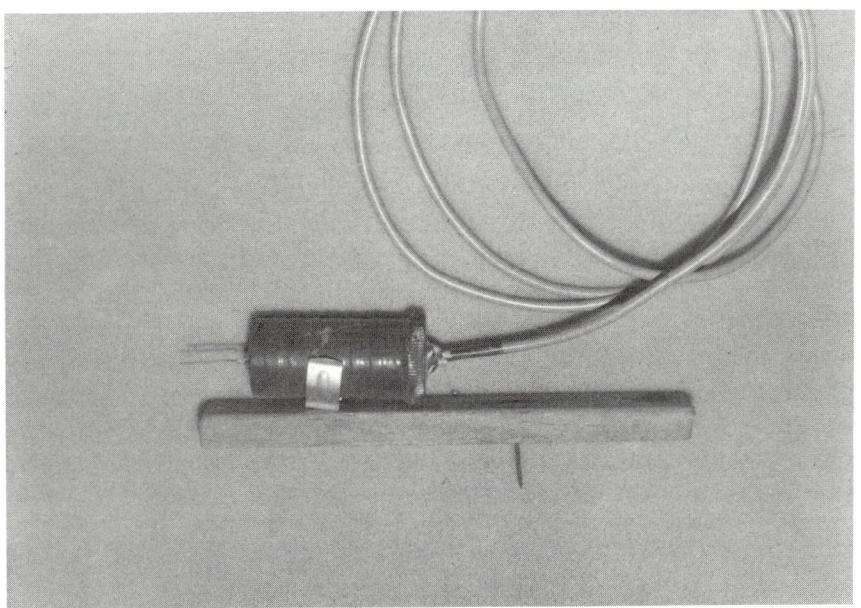

FIGURE 5—1F: MOISTURE-WATER DETECTOR PROBE ASSEMBLY

might want to aerosol spray the probe assembly with a good coat of epoxy enamel.

How It Works

Please refer to Figure 5—1A for the following explanation. S1 applies power to the on-off pilot lamp and to transformer T1. The detector is now in the standby condition until moisture or water is sensed across the moisture sensing prods. When moisture is detected the SCR turns on, energizing the relay coil and activating A1, the piezoelectronic alarm. The normally open relay contacts close, applying power to the moisture controlled outlet, SO1. Neon lamp, L2, will also light showing that the outlet is activated. As soon as the moisture sensing prods become dry the SCR will turn off and the detector will automatically return to the standby condition.

Manual override switch S3 applies power to the moisture controlled outlet regardless of the state of the sensing prods. S2 can be switched open to kill the alarm once it has sounded its initial warning. C1 holds the relay energized during the negative half-cycle of the AC alternation.

Operating Tips

Position the moisture sensing prods wherever moisture or water is to be detected. However, one note of caution on probe placement is that the moisture sensing prods are extremely sensitive. For example, suppose you are monitoring the low point of a concrete basement floor for flooding. If the prods are actually touching the floor there's a good chance the detector will become activated just from surface moisture during damp weather. To prevent this type of false alarm it's a good idea to always keep the prods 1/32" or so away from contacting anything. In this way the floor would have to be wet before the detector becomes operational.

5—2: SLEEP INDUCER

What It Does

Are you having trouble falling asleep at night? Here's a little gem that will help you sleep like a baby.

Turn the sleep inducer on to hear a broadband white noise. You can make it super soft or plenty loud to fit your fancy. Some people say that white noise reminds them of falling rain; others say it's like wind through the trees. In any event, the sound is a good distracter and will effectively mask out house and neighborhood noises from your bedroom.

In addition to the white noise feature, the sleep inducer also displays a random pattern of blinking LEDs. Ever try counting sheep to fall asleep? Now you can count LED blinks and before you know it you'll be in dreamland.

Here's the Schematic

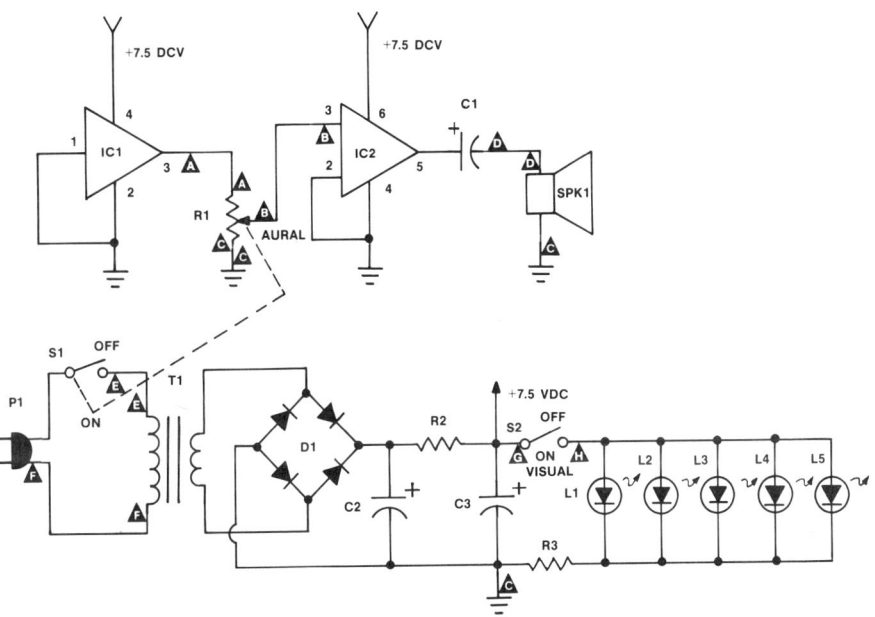

FIGURE 5—2A: SLEEP INDUCER SCHEMATIC

Parts to Use

C1 - 250 MF, 15 V	R1 - 10K potentiometer with switch
C2 - 1000 MF, 15 V	
C3 - 4 MF, 15 V	R2 - 47 ohm ½W, ±10%
D1 - 1A, 50 PIV bridge	R3 - 330 ohm 1W, ±10%
IC1 – MM5837 noise source	S1 - SPST switch (part of R1)
IC2 - LM386 audio amplifier	S2 - SPST switch
L1-L5 - LED flasher	SPK1 - 8 ohm, 2″ speaker
P1 – AC plug	T1 - 117:6.3 VAC, 300 MA

Misc.- enclosure, hardware, knob, LED clips, line cord, PC board, strain relief, etc.

The PC Layout You'll Need

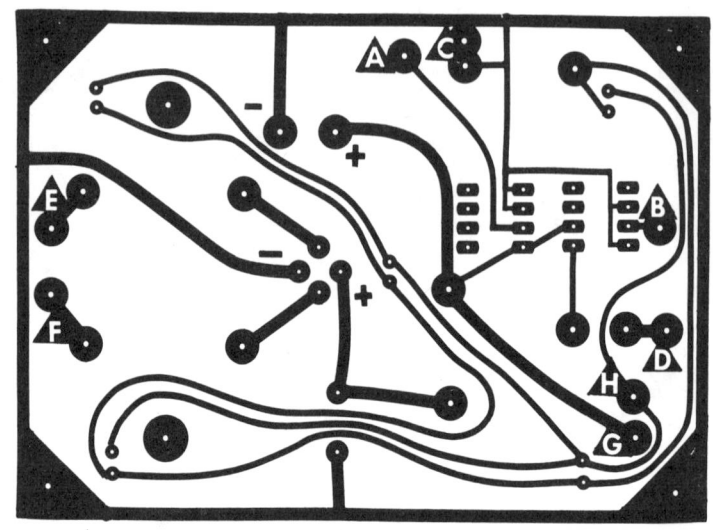

FIGURE 5—2B: SLEEP INDUCER PC LAYOUT (ACTUAL SIZE)

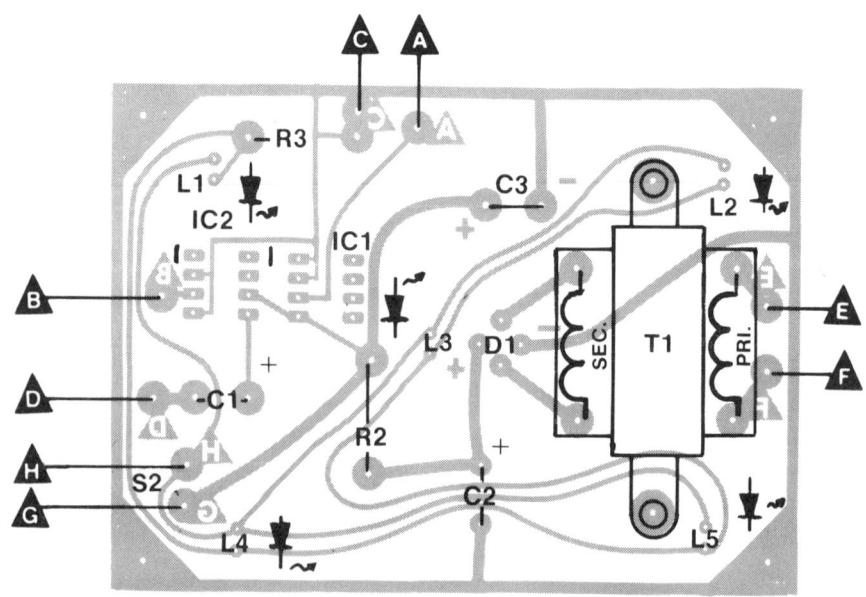

FIGURE 5—2C: SLEEP INDUCER COMPONENT PLACEMENT GUIDE (TOP VIEW)

Helpful Construction Hints

A small plastic box or instrument case makes an ideal enclosure. You might want to put contact paper over the outside to match your decor.

All parts are mounted on the PC board except for the switches, potentiometer, speaker, and possibly the LEDs. A neat way to mount the flashing LEDs is from the foil side of the board. The board can then be placed in the enclosure with the part side down, the foil side and the five LEDs up. Use a piece of translucent or transparent plastic for the front cover.

Of course, the LEDs can be individually mounted on the front panel in the conventional manner using LED clips. Just run the connecting wires between the LED terminals and the appropriate PC foil islands.

How It Works

Please see Figure 5—2A for the following explanation. Line voltage is transformed to 6.3 VAC by T1. Full wave bridge rectifier D1 converts the AC to DC. Filtering is provided by C2, R2 and C3. The DC voltage is used to power each IC and the flashing LEDs.

IC1 is a digital white noise generator that provides uniform noise quality and amplitude. The white noise is tapped off potentiometer R1 and fed to audio amplifier IC2. C1 couples the amplified white noise to a small speaker (SPK1).

LEDs L1-L5 are self flashing. Since all the LEDs share a common dropping resistor, unusual flashing patterns develop which make a novel display. S2 switches off the LEDs for aural only operation.

Operating Tips

The sleep inducer is designed so that you have three types of options:

1. White noise only
2. Flashing LEDs only
3. Both white noise and flashing LEDs

The loudness of the white noise can be controlled by potentiometer R1. The tone can also be changed somewhat by how the speaker is directed. It can be pointed directly at you, away from you or from side to side. Even setting the sleep inducer with the speaker side down will create a different kind of sound. Choose the sound level and tone most pleasing to you.

Pleasant dreams!

5—3: STEREO SYSTEM GENIE

What It Does

If you are frequently lulled to sleep by your favorite music, you need the stereo system genie. Eliminate the energy waste and the risk of overheating and the possible component damage that can occur if your stereo system is on all night. This project requires no complicated internal connections to your turntable or amplifier; it automatically turns off your stereo system when the last record has played. Why not relax, enjoy the music, and let this easy-to-build project work for you?

Here's the Schematic

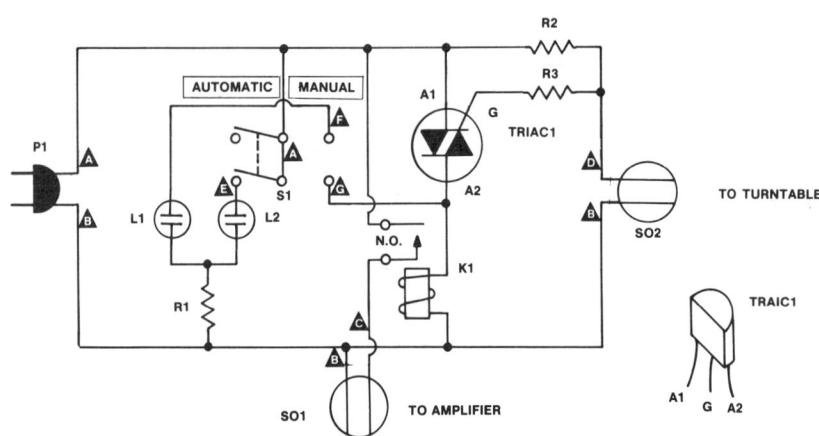

FIGURE 5—3A: STEREO SYSTEM GENIE

Parts to Use

K1 - 110 VAC SPST   R3 – 47 ohm, ½ watt, 10%
  relay 10 a. contacts   S1 - DPDT switch
  (Radio Shack 275-217)   SO1, SO2 - 110 VAC outlets
L1, L2 - NE-2 neon lights TRIAC 1 – sensitive gate,
P1 - 110 VAC plug   200 volt, (GE SC92D)
R1 – 56 K ohm, ½ watt, 10%
R2 – 27 ohm, 1 watt, 10%

Misc.-enclosure, grommet, hardware, hookup wire, lamp
  lenses, line cord, PC board, etc.

The PC Layout You'll Need

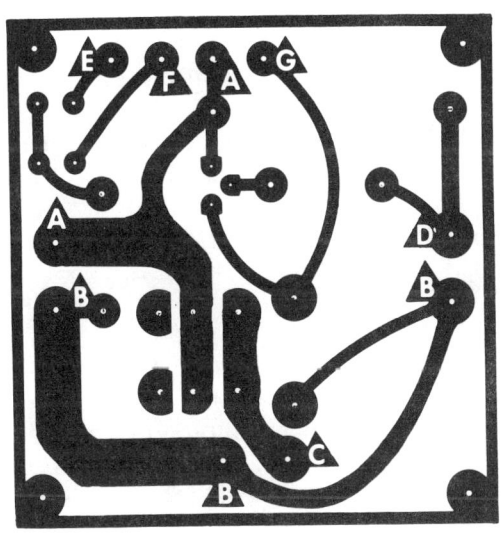

FIGURE 5—3B: STEREO SYSTEM GENIE PC LAYOUT
(ACTUAL SIZE)

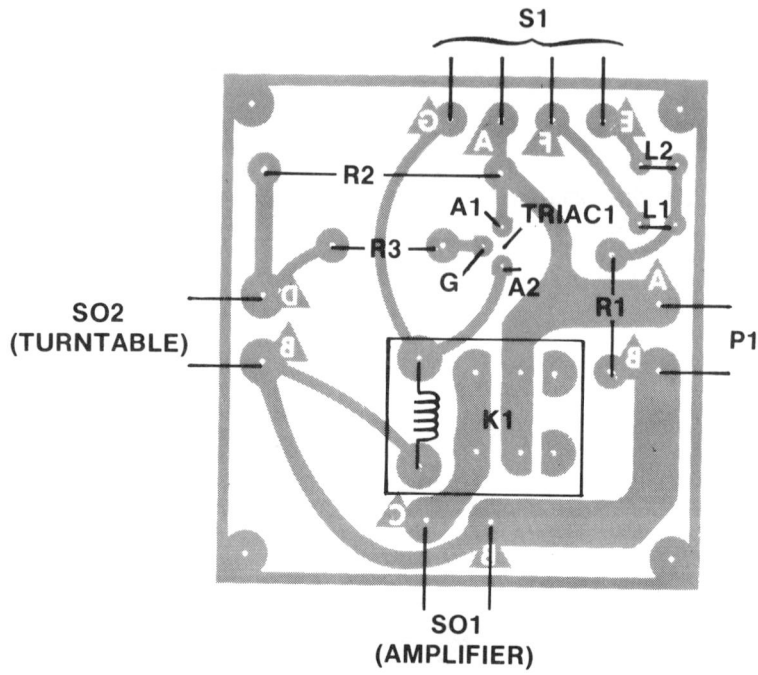

FIGURE 5—3C: STEREO SYSTEM GENIE COMPONENT PLACEMENT GUIDE (TOP VIEW)

Helpful Construction Hints

Insert and solder the component parts in the printed circuit board (see Figure 5—3C) carefully observing the lead placement of the triac. If a relay other than the one specified is used, be certain the normally open contacts are utilized. Mount the switches, lamp lenses, and printed circuit in a minibox. For convenience, SO1 and SO2 may be installed in the rear apron of the enclosure. Since this is a line operated device, all circuitry must be electrically isolated from the enclosure. Verify this isolation with an ohmmeter. Infinite resistance should be measured between the minibox and any terminal connections with the unit.

How It Works

When a turntable connected to SO2 is operating, a voltage drop is created across R2. This 2-3 volt drop triggers the triac into conduction, thus causing the relay coil to be energized. The closed relay contacts now supply power to the externally connected amplifier. Turning off the turntable or record changer removes the current flow through R2 and the subsequent voltage drop. The triac ceases conduction, because of the lack of potential voltage between the gate and anode 1. The relay is now de-energized and power is removed from the amplifier system (SO1). L1 and L2 indicate the mode of operation as selected by S1. R3 provides current limiting to the triac's gate.

Operating Tips

Plug the turntable and amplifier into outlets SO2 and SO1 respectively. Position S1 in the automatic mode and apply line voltage to the genie. When the turntable or record changer is actuated, power will be supplied to the amplifier. After completion of the playing cycle, the amplifier will turn off. Toggling S1 to the manual position permits continuous amplifier operation, as indicated by illumination of L2.

Tape decks and recorders incorporating "end of tape shut-off" can also be used for sensing control. However, because of their differing wattage requirements, the value of R2 must be changed.

Warning: Do not connect a device requiring more than 10-15 watts to SO2, without first changing the value of R2. (Refer to Figure 5—3D.)

Wattage rating of device connected to SO2	Value of R2 required (tolerance ±10%)
10-15 watts	27 Ohm, 1 watt
30 watts	10 Ohm, 1 watt
50 watts	6.8 Ohm, 2 watt
75 watts	4.7 Ohm, 2 watt
100 watts	3.3 Ohm, 5 watt
120 watts	2.7 Ohm, 5 watt
150 watts	2.2 Ohm, 5 watt

FIGURE 5—3D: R2 VALUE CHART

5—4: TOUCH SWITCH

What It Does

Picture yourself coming home from a hard day's work and slumping down in your easy chair. You pick up the paper and begin to read the headlines. It's too dark; you could use some light. On the end table next to your chair is a strange little box. Your hand reaches over and ever so lightly touches the box. Immediately, the table lamp comes to life flooding your reading area with light. After you finish reading the paper, a little snooze is in order. Again your hand drifts over effortlessly to the box. At contact the table lamp immediately turns off leaving the room in soothing darkness.

The previous paragraph illustrates only one use for the touch switch. Your imagination is all that limits the myriad applications for this magic little box. The outside ingredients of the box consist of a touchplate, an AC receptacle, and an AC cord. The AC cord is inserted into a standard AC outlet. Whatever is plugged into the touch switch box is controllable both on and off with the gentlest tap. It's the greatest!

Here's the Schematic

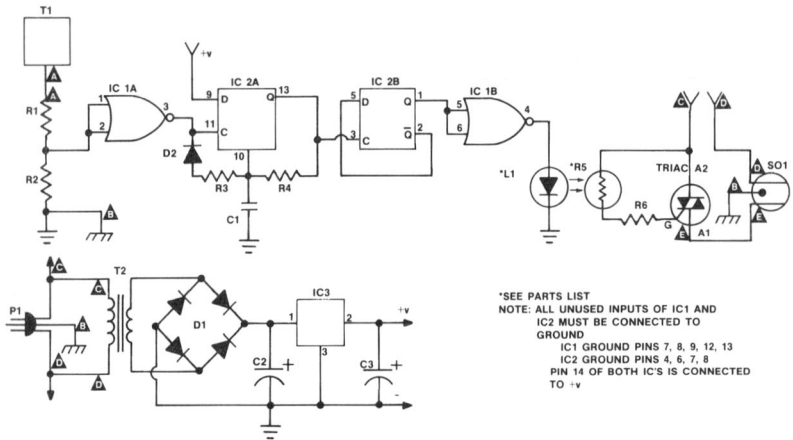

FIGURE 5—4A: TOUCH SWITCH SCHEMATIC

Parts to Use

C1 - .1 MF, 50 V
C2 - 1000 MF, 15 V
C3 - 2 MF, 15 V
D1 - 1A, 100 PIV bridge
D2 - 1N4148 silicon diode
IC1A, B - 4001B CMOS
    gate
IC2A, B - 4013B CMOS
    flip-flop
IC3 - 7805 voltage
    regulator, 5 V
*L1 - LED, clear
P1 - AC plug with safety
    ground
R1 - 2.2 M ½W, ±10%

R2 - 60 M ½W, ±10%
R3 - 3.3 K ½W, ±10%
R4 - 510 K ½W, ±10%
*R5 - CDS photoconductor,
    ¼" diameter, dark
    resistance 100 K+, light
    resistance 50 ohms
R6 - 4.7 K ½W, ±10%
SO1 - AC receptacle
T1 – 1" × 1" touchplate,
    aluminum
T2 - 117:6.3, 200 MA
    transformer
Triac - 15 A, 400 PIV triac

Misc.- enclosure, hardware, line cord, PC board, strain
    relief, etc.
*NOTE: L1 and R5 can be substituted by LED - photo-
    conductor opto-isolator.

The PC Layout You'll Need

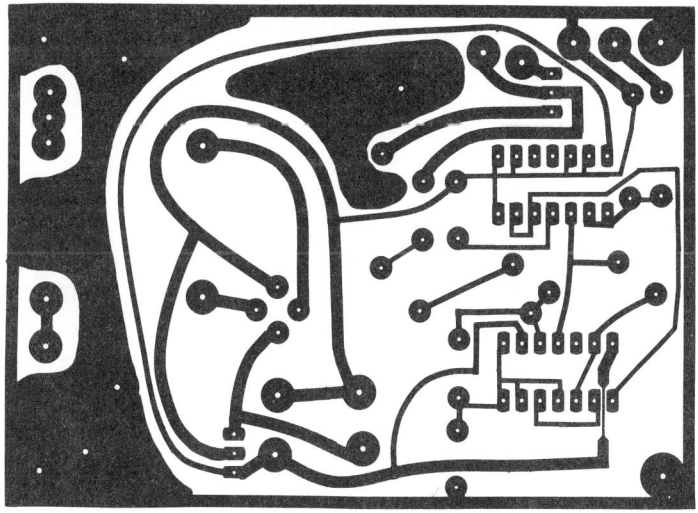

**FIGURE 5—4B: TOUCH SWITCH PC LAYOUT (ACTUAL
SIZE)**

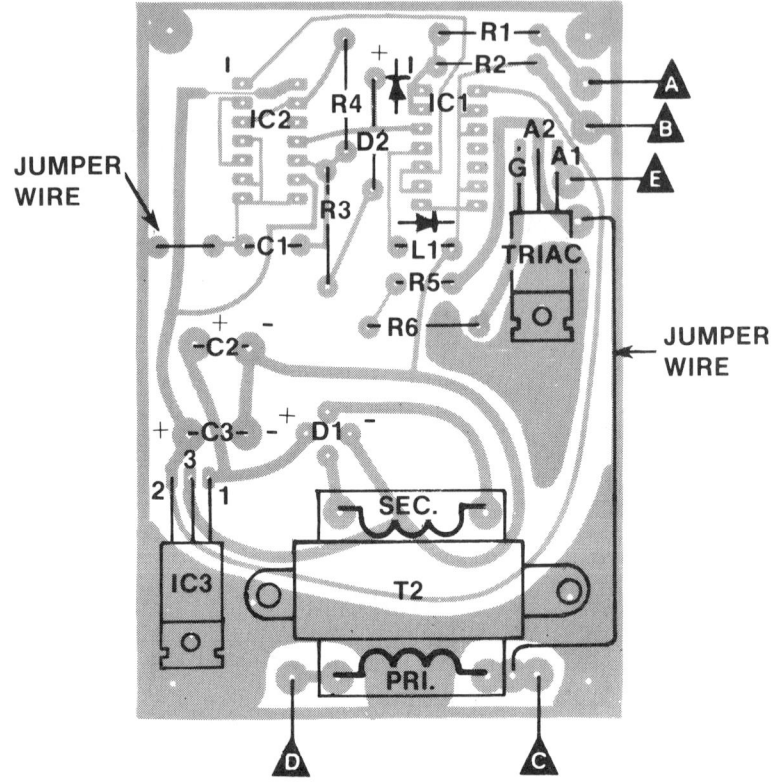

**FIGURE 5—4C: TOUCH SWITCH COMPONENT
PLACEMENT GUIDE (TOP VIEW)**

Helpful Construction Hints

Construction is fast and easy once the PC board is made. (Refer to Section 1—2.) Although the schematic appears to be quite complicated, the actual building goes rather rapidly.

Mount the PC board in a small opaque plastic or bakelite case. The aluminum touchplate can be installed in a convenient place on the top.

Be sure to heat sink the triac. Use silicon grease between the triac case and the heat sink for good thermal conductivity. The triac's case or tab will probably not be insulated from the anode so be certain that the heat sink is electrically insulated from the rest of the project.

L1 and R5 (see Figure 5—4A) can be conveniently replaced with an LED photoconductor opto-isolator. A separate LED and photoresistor work just as well if you have trouble finding an opto-isolator locally. The other part that might be hard to find is the 60 meg ohm resistor (R2). This resistor value is not critical except that it should be an extremely high resistance. If you can't locate this resistor, series connect a few of the highest meg ohm resistors available to approximate the value.

How It Works

As your finger touches or comes close to the touchplate, 60 HZ power line hum is induced into the MOSFET gate IC1A. (See Figure 5—4A.) The hum is cleaned up by the gate and used to trigger flip-flop IC2A. A second flip-flop (IC2B) is necessary to keep the output signal up or down when your finger is removed from the touchplate. IC1B drives LED L1 which is physically positioned to shine on photoresistor R5. When L1 light strikes the photoresistor, its resistance decreases causing gate current to flow in the triac. The triac fires switchig on current to socket S1 and powering whatever is plugged into it.

In the power supply, line voltage is transformed to 6.3 VAC by T2. The bridge rectifier changes the AC to DC which is then filtered by C2 and C3. IC3, a 5-volt regulator, keeps the output at a constant 5-volt level.

Operating Tips

The touchplate works best when the project ground is connected directly to the safety ground of the house electrical system as shown in the schematic, Figure 5—4A. In some applications the safety ground connection may not be needed.

If the touchplate does not seem to work, try reversing the AC plug in the receptacle. It will work only when the plug is one way. Also, the case must be lightproof so that room light does not interfere with the touchplate photoconductor operation.

5—5: TV REMOTE SOUND CONTROL

What It Does

Many people would often like to enjoy TV without disturbing other folks in the room. The TV remote sound control is the answer.

(See Figure 5—5A.) Just plug an earphone or headset into the remote sound control box and flick a switch. The TV sound will shut off. You can adjust the phone volume to suit your hearing preference. This little device is just the ticket for people who are hard of hearing.

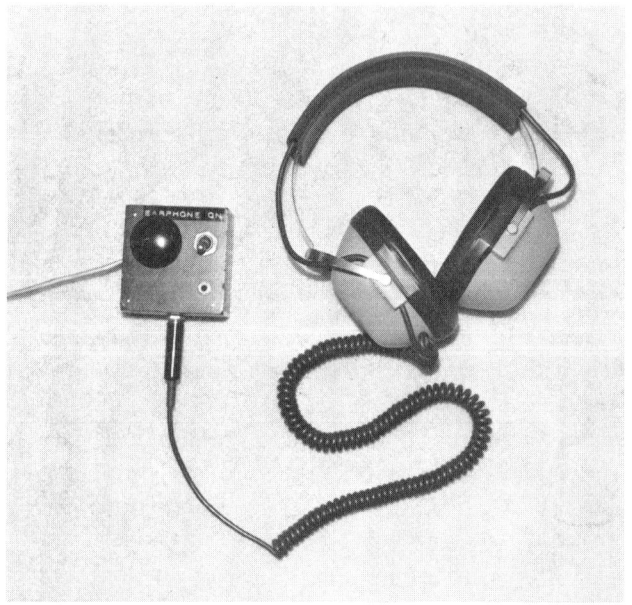

FIGURE 5—5A: TV REMOTE SOUND CONTROL

Now you don't need to have the TV blaring. Keep the sound at normal volume; the hard of hearing person can use the TV remote control to hear better than ever.

Another feature of this neat little device is the ability to kill TV sound from the comfort of your easy chair. Even when you are listening to TV normally why not click off those annoying commercials? It's as easy as flicking a switch.

Here's the Schematic

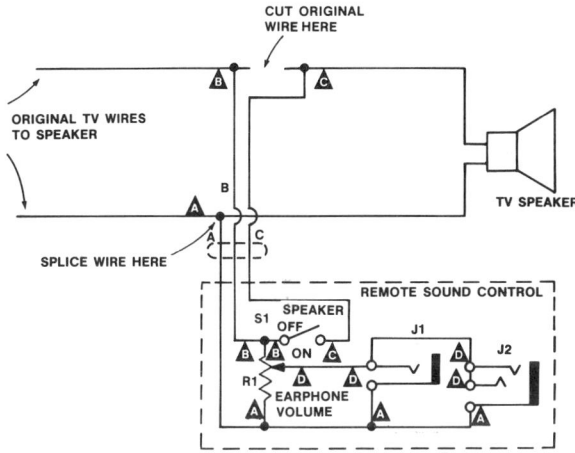

FIGURE 5—5B: TV REMOTE SOUND CONTROL SCHEMATIC

Parts to Use

J1 – phone jack, mono   R1 – potentiometer, audio
J2 – phone jack, stereo   taper
                        S1 – SPST switch
Misc.- cable (three-conductor stranded, 20'), enclosure,
        grommet, knob, wire nuts, etc.

Helpful Construction Hints

Monophonic and stereophonic phone jacks are used in this project to accommodate both types of headphones. Select the type of jack to fit the plug of any headphones you might already have. If you use a stereo jack be sure to parallel wire both inside J2 jack connections as shown in Figure 5—5B so each side of the headset receives sound.

Use three-conductor stranded wire cable between the TV and the remote sound control box. Allow plenty of length so that all viewing areas can be easily reached.

The quickest and easiest way of connecting the remote control cable to the TV speaker wires is with wire nuts. Open the back of the TV looking for the speaker. Find the two wires going to the speaker terminals. Cut one of them. Strip the insulation from both pieces. Wire-nut control cable wire C to the cut wire section that is still connected to the speaker. (Refer to Figure 5—5B.) Wire-nut control cable wire B to the other cut section. Now cut the other original speaker wire stripping insulation from both pieces. Wire-nut cable wire A to both of the pieces just cut. That's it. Pop the back on the TV. The remote sound control is ready for action.

How It Works

The beauty of this circuit lies in its simplicity coupled with the fact that it is so quick and easy to connect to any TV set.

Refer to Figure 5—5B for the following explanation. The speaker on-off switch (S1) opens or closes the circuit to the TV speaker from a remote location. The audio is tapped off from the speaker wires by the earphone volume control (R1). Since R1 is wired independently from the switch circuit, the phone volume can be adjusted at all times regardless of whether the speaker is on or off.

The arm of R1 feeds phone jacks J1 and J2. J1 will accept a mono-type plug and J2 takes a stereo plug. J2 is wired so that both sides of the stereo headphones will receive audio.

Operating Tips

The TV remote sound control will not interfere in any way with normal TV reception. The only possible problem you might encounter is if someone inadvertently left the speaker switch in the off position. A flick of the switch will quickly restore normal TV operation.

If more than one person uses the TV remote sound control, it's a good idea to always turn the earphone volume control to off after using it. The next person who wears the headphones can then adjust the volume to his own listening level. Remember, what one person hears as normal someone else may hear as thunder!

A person who is really hard of hearing may have to turn the TV volume up, switch off the TV speaker, and adjust the earphone volume control to maximum. However most people find that the TV volume control can remain in the normal listening position.

5—6: PHOTORESISTIVE SENSING UNIT

What It Does

Here is a project that offers you safety, convenience, and protection in a single unit. With the photoresistive sensing unit, you can automatically turn your lights on at dusk and off at dawn. You can discourage prowlers by giving your home a "lived-in" look when you are away. A simple flick of a switch allows the sensing unit to power appliances in the presence of light—great for starting the coffee pot in the morning. Build the photoresistive sensing unit and let electronics make your life easier.

Here's the Schematic

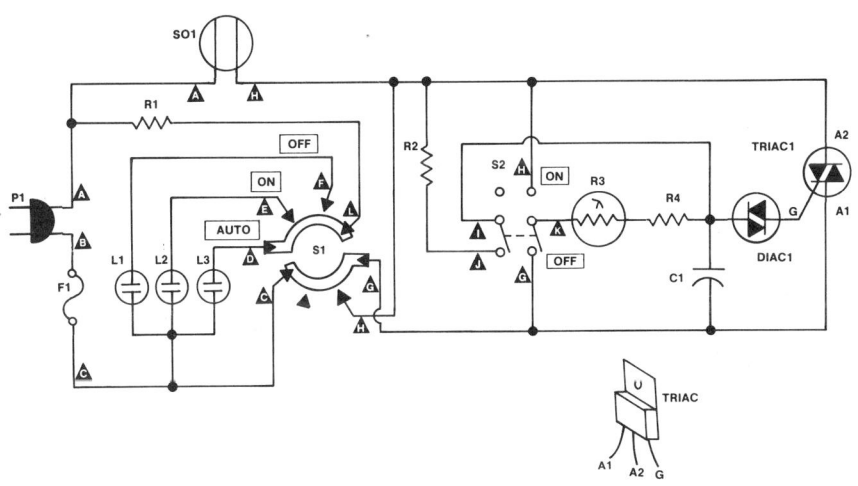

FIGURE 5—6A: PHOTORESISTIVE SENSING UNIT

Parts to Use

C1 – .1 mfd. 200 volt	L1, L2, L3 – NE-2 neon
DIAC1 – trigger diac	lights
28-32 volt	P1 – 110 VAC plug
F1 – 6 a. fuse	R1 – 56 K ohm, ½ watt, 10%

R2 – 15 K ohm, 2 watt, 10%   S1 – 2 pole, 3 position
R3 – cadmium sulfide   rotary switch
        photoresistor (Radio   S2 – DPDT switch
        Shack 276-116)   SO1 – 110 VAC outlet
R4 – 1 K ohm, ½ watt, 10%   TRIAC 1 - 6 a. 200 volt
                                                  (Radio Shack
                                                  276-1001)

Misc.- enclosure, fuse holder, hardware, hookup
        wire, knob, lamp lenses, line cord, PC board, etc.

The PC Layout You'll Need

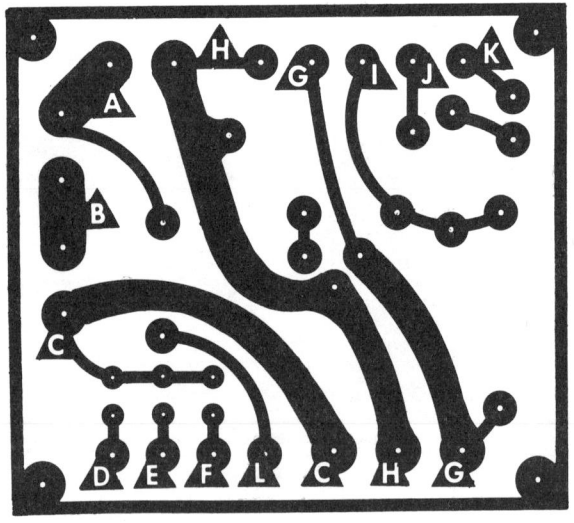

**FIGURE 5—6B: PHOTORESISTIVE SENSING UNIT PC
LAYOUT (ACTUAL SIZE)**

Helpful Construction Hints

Insert and solder the component parts in the PC board, carefully noting the lead placement of the triac. (Refer to Figure 5—6C.) Protect the lamp and photocell leads with insulating spaghetti. To utilize the full power handling capacity of the triac, a suitable heat sink should be attached to the mounting tab. When installing the neon lamps in their respective lenses, avoid sharp bends in the

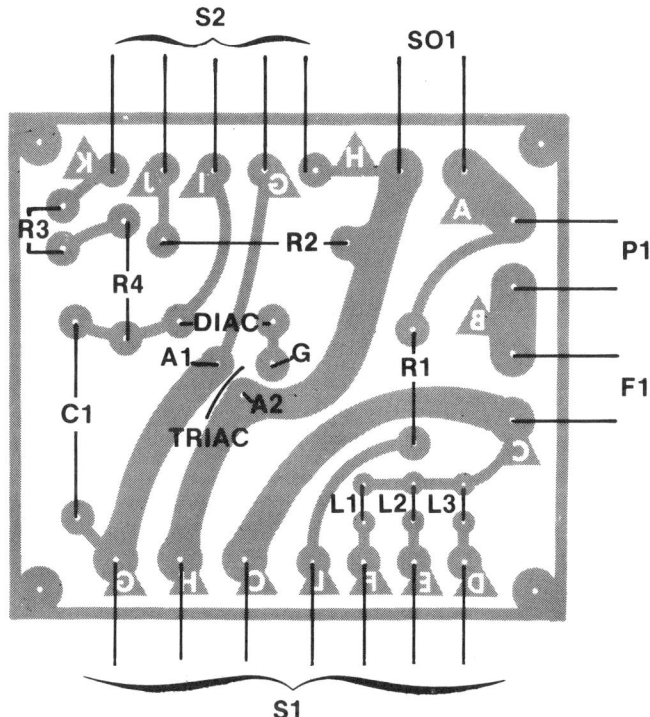

FIGURE 5—6C: PHOTORESISTIVE SENSING UNIT COMPONENT PLACEMENT GUIDE (TOP VIEW)

fragile leads, particularly where they enter the glass bulb. To prevent unwanted ambient light from reaching the photocell, a cylindrical light shield may be attached to R3. Since this is a line operated device, be sure that all circuitry is electrically isolated from the enclosure before operating the sensing unit. Ohmmeter checks between the minibox and circuit connections should indicate infinite resistance, thus providing safe and reliable operation.

How It Works (See Figure 5—6A)

S1 selects the mode of operation as indicated by L1, L2, and L3. When S1 is in the "auto" position, photocell R3 controls the power available to SO1 via the triac. The position of S2 determines if power is applied to the externally connected device in the presence or absence of light. With S2 "off" and the photocell exposed to light, the low resistance of R3 prevents C1 from reaching the breakdown voltage of the diac. Since the triac requires a gate pulse from diac1 to conduct, the series connected SO1 will not receive power. Removing the light source permits C1 to "fire" the diac, thus allowing the triac to turn on.

Toggling S2 "on," with the photocell exposed to light, permits C1 to sufficiently charge and trigger the triac. Extinguishing the light source increases R3's resistance, thereby preventing the triac from conducting, and removing power to SO1.

Positioning S1 "on" provides continuous power to the outlet, as indicated by illumination of L1. Selection of the "off" position for S1 disables SO1.

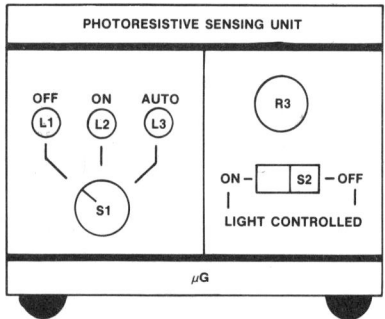

FIGURE 5—6D: PHOTORESISTIVE SENSING UNIT (FRONT PANEL VIEW)

Operating Tips

For light-controlled applications, select the "auto" mode for S1, and choose either the light-activated "on" or "off" position for S2. (See Figure 5—6D.) With S2 "on," power will be applied to SO1 only when the photocell is exposed to light. Extinguishing the light source will turn off an externally connected device. Toggling S2 "off" will apply external circuit power only in the absence of light. Turn S1 "on" if continuous power to SO1 is desired. To disable the receptacle, select the "off" position for S1. Uses of the control unit include entrance warning systems and darkness-actuated lights in the home.

5—7: TOOL MAGNETIZER AND DEMAGNETIZER

What It Does

Here is a handy one-evening project that allows you to quickly magnetize or demagnetize your workbench tools. Simply press a button and that ordinary screwdriver becomes magnetic—a real timesaver when installing or retrieving nuts, screws, or washers. Press the button again, and magically the screwdriver loses its magnetism. Why not build the tool magnetizer and demagnetizer today? You will find it an indispensable additon to your workshop equipment.

Here's the Schematic

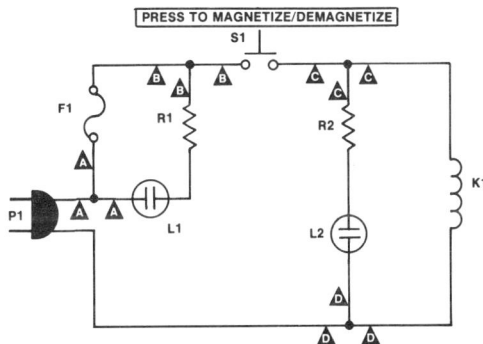

FIGURE 5—7A: TOOL MAGNETIZER AND DEMAGNETIZER

Parts to Use

F1 – 2 a. fuse	P1 – 110 VAC plug
K1 – 110 VAC solenoid	R1, R2 – 56 K ohm,
(see text)	½ watt, 10%
L1, L2 – NE-2 neon	S1 – SPST push button switch
lights	(spring loaded)
Misc. - enclosure, fuse holder, grommet, hardware, hookup wire, lamp lenses, line cord, terminal strip, etc.	

Helpful Construction Hints

Mount the switch, lamp lenses, solenoid, fuse holder, etc. in a small enclosure. Provide a ½″ opening in the enclosure to allow access to the field coil. Point-to-point wiring is suggested, with a terminal strip provided for the indicator lamps and their respective current limiting resistors. Any 110 VAC solenoid may be used for K1. Before installing K1, remove and discard the solenoid armature assembly. Since this is a line operated device, be certain that all circuitry is electrically isolated from the enclosure.

How It Works (refer to Figure 5—7A)

Closing S1 applies an alternating current to K1. This alternating current produces a changing magnetic field within the solenoid coil. If an iron or steel tool is placed within this field and power is removed, the tool will become magnetized. However, if the tool is removed from the magnetic field when K1 is still energized, the tool will encounter a progressively weaker and reversing magnetic field, and become demagnetized. F1 provides circuit protection, and L1 and L2, respectively, serve as "open fuse" and "power on" indicators.

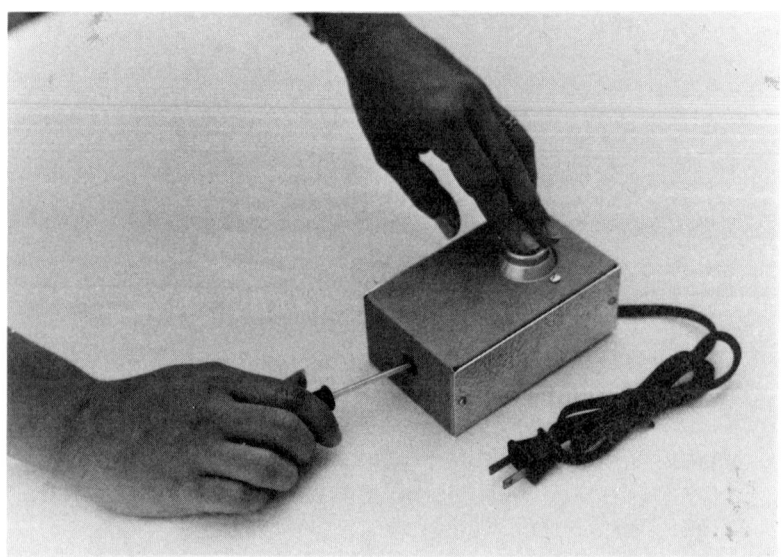

FIGURE 5—7B: MAGNETIZING A SCREWDRIVER

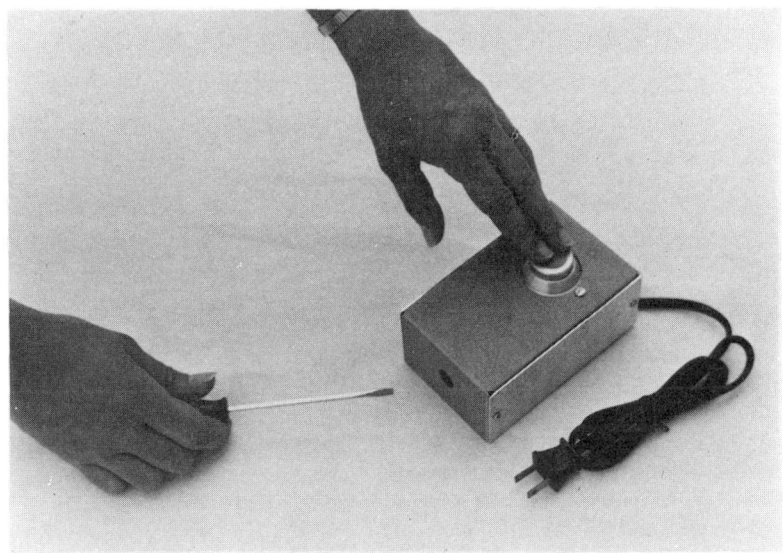

FIGURE 5—7C: DEMAGNETIZING A SCREWDRIVER

Operating Tips

Magnetizing: Insert the tool to be magnetized into the field coil and momentarily depress S1 (Figure 5—7B). Then, release S1 and remove the magnetized tool.

Demagnetizing: Insert the tool into the field coil while depressing and holding S1. Slowly remove the tool with K1 still energized (Figure 5—7C), and release S1. *Note*: Do not depress S1 longer than 5 seconds, or overheating of K1 may result.

5—8: POWER FAILURE LANTERN

What It Does

Here is a handy project that will provide you with automatic emergency lighting in the event of a power outage. Similar to units found in retail establishments, the power failure lantern offers greater versatility by doubling as stationary standby lighting or a portable lantern floodlight within easy reach. As it incorporates rechargeable batteries, this lantern is always ready when the lights go out. Why wait to be left in the dark? Enjoy the safety and convenience of the power failure lantern today.

FIGURE 5—8A: POWER FAILURE LANTERN (FRONT VIEW)

Here's the Schematic

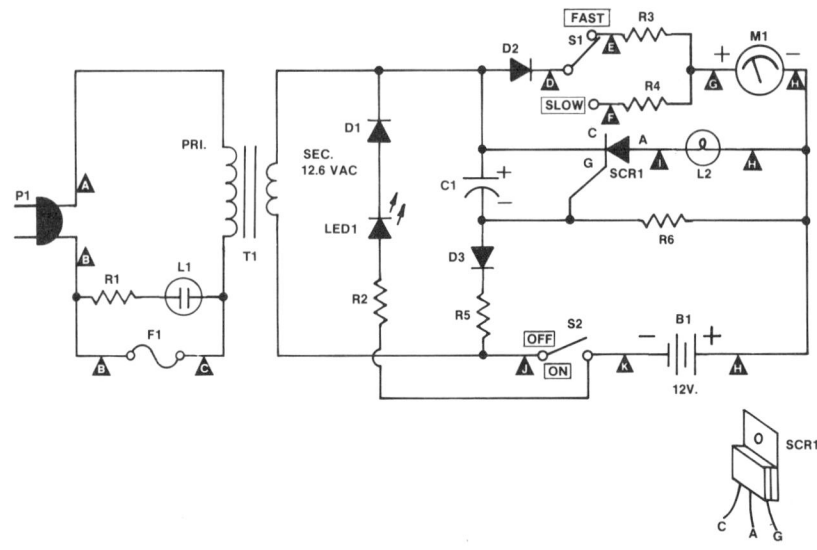

FIGURE 5—8B: POWER FAILURE LANTERN

Parts to Use

B1 – 12 volt rechargeable battery (see text)
C1 – 200 mfd, 25 volt
D1, D2, D3 – 1A, 50 PIV   R3 – 5 ohm, 5 watt, 10%
F1 – ½ ampere fuse   R4 – 27 ohm, 2 watt, 10%
L1 – NE-2 neon light   R5 – 47 ohm, 1 watt, 10%
L2 – 12 volt light   R6 – 1.5 K ohm, ½ watt, 10%
    (see text)   S1 - SPDT switch
LED1 – jumbo red   S2 – SPST switch
M1 – 0 - 1 A. ammeter   SCR1 – 4A, 50 PIV,
P1 – 110 VAC plug   GATE: .8 volt .2 ma.
                                (GE C106F1)
R1 – 56 K ohm, ½ watt, 10% T1 – 12.6 volt, 1.2 ampere
R2 – 470 ohm, ½ watt, 10%   transformer (Radio
                                Shack 273-1505)

Misc. - assorted hardware, enclosure, fuse holder,
    hookup wire, knob, line cord, etc.

The PC Layout You'll Need

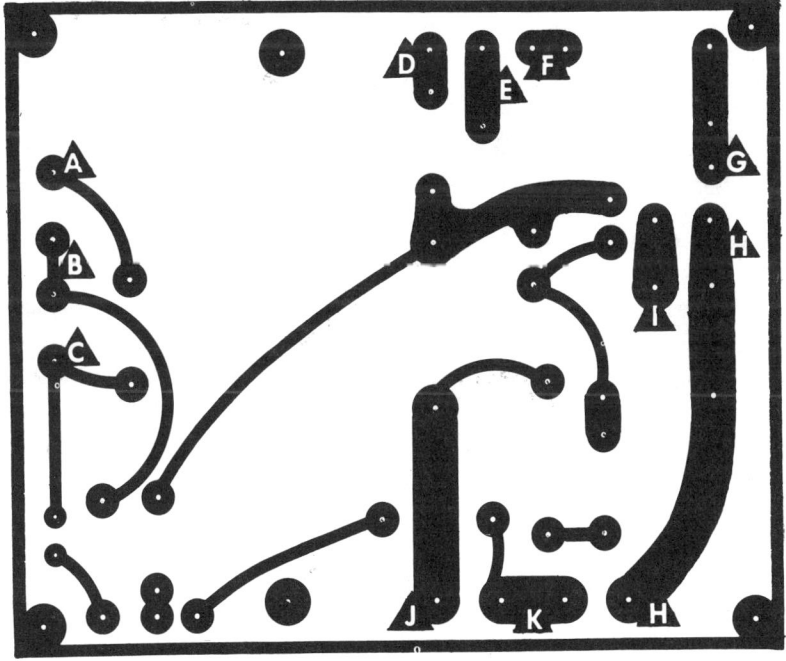

**FIGURE 5— 8C: POWER FAILURE LANTERN PC LAYOUT
(ACTUAL SIZE)**

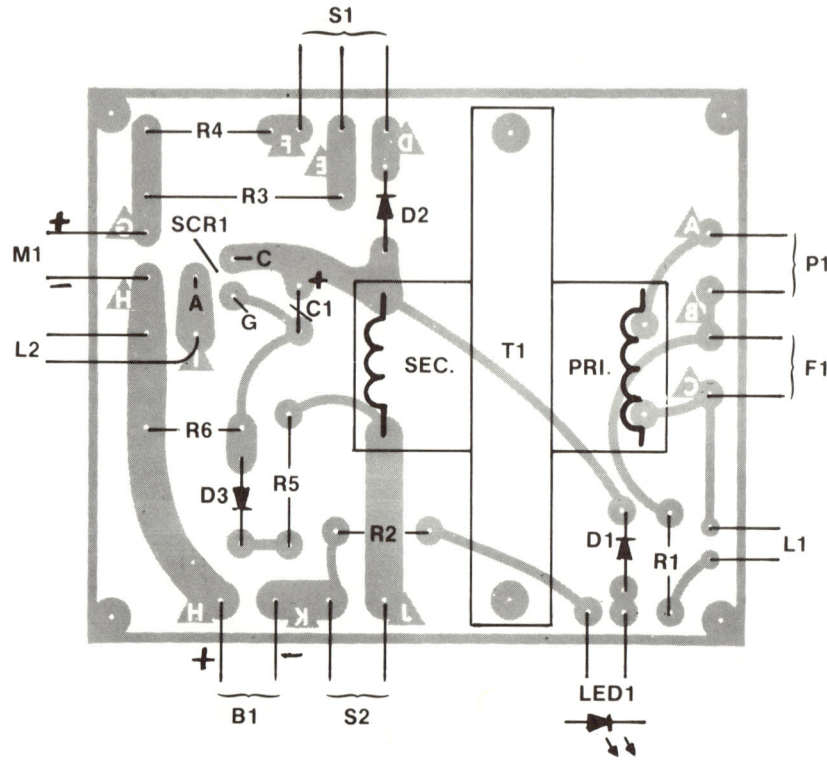

**FIGURE 5—8D: POWER FAILURE LANTERN PLACEMENT
GUIDE (TOP VIEW)**

Helpful Construction Hints

Mount the component parts on the printed circuit board (see Figure 5—8D) carefully observing the polarities, of the capacitor, diodes, and LED. To utilize the maximum current rating of the SCR, provide a suitable heat sink for the anode tab. Because this is a line operated device, the transformer's primary circuitry must be electrically isolated from the enclosure. The integrity of this circuit can be verified using an ohmmeter. Infinite resistance should be measured between the enclosure and any of the primary terminals.

Any rechargeable battery may be used for B1. Nickel-cadmium or gel-cell packs are recommended, although small lead-acid (motorcycle) batteries may be used. If lead-acid batteries are selected, provide small ventilation holes in the enclosure to allow for escaping vapors.

The values of R3 and R4 may be modified to best suit the charging requirements of the battery used. Lower ohmic values provide a fast charge, while higher resistance results in a slower charging rate.

A discarded lantern reflector with a separate 12-volt bulb makes an excellent assembly for L2. If increased illumination is desired, a small sealed-beam unit may be used. However, the battery's charge will be more rapidly depleted. Be sure that the current rating of the SCR is not exceeded by L2.

How It Works (Refer to Figure 5—8B)

L1, with the series current limiting resistor R1, serves as an "open" fuse indicator, and normally is not illuminated. When AC power is applied and S2 is closed, B1 is charged by the half-wave rectifier circuit consisting of D2, and either R3 or R4. S1 selects the rate of charge, with M1 serving as a current monitor.

C1 charges via D3 and R5, placing a negative voltage on the SCR's gate. This negative voltage overrides the positive potential from the battery through R5. In addition, C1 places a positive charge on the SCR's cathode, thus turning "off" the SCR. LED1 receives power from T1, but only if S2 is closed. R2 provides current limiting and D1 increases the LED's PIV rating.

When the power failure lantern is unplugged, or a power outage occurs, the charge on C1 dissipates. This allows a positive potential to be placed on the SCR's gate and a negative potential on the cathode from B1. The SCR is now "turned on" and permits current to flow from B1, through the transformer's secondary, lighting L1.

Operating Tips

Plug the unit into a convenient wall outlet. Close S2 while observing the illumination of LED1. Select either the slow or fast charge position for S1. This selection will depend upon the battery's state of charge. M1 monitors the charging current, and will indicate a decreased reading as B1 develops a full charge. When B1 is fully charged, remove the AC line power. L2 will now light, and the unit can be used as a lantern flashlight. Open S2 to turn off L2. S2 must be closed to recharge B1, and for automatic standby lighting. If S2 is open when the system is returned to AC power, LED1 will not light.

5—9: INTERCOM SYSTEM

What It Does

Here is a convenience project that is easy and inexpensive to build. Save time and steps by building an intercom system for your home. With one master and one remote unit, you can communicate from opposite ends of your home without raising your voice. Talk between kitchen and basement or house and garage with ease. So, stop shouting and make you life easier by building the home intercom system.

Here's the Schematic

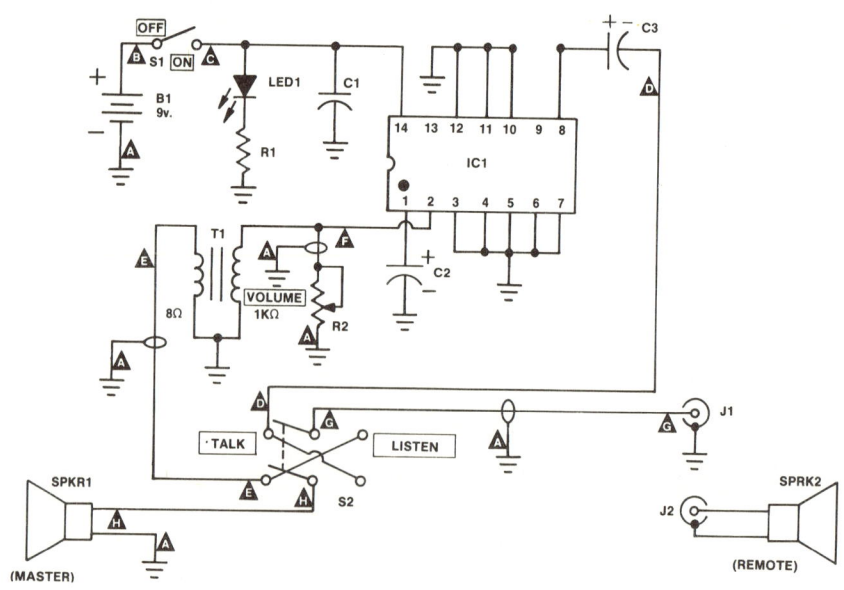

FIGURE 5—9A: INTERCOM SYSTEM

Parts to Use

B1 – battery (see text)   R1 – 270 ohm, ½ watt, 10%
C1 – .1 mfd. 25 volt disc. R2 – 10 K ohm audio taper
C2 – 47 mfd. 15 volt   S1 – SPST switch
C3 – 470 mfd. 15 volt   S2 – DPDT switch
IC1 – LM380N, 2 watt   SPKR1,SPKR2 – 8 ohm, 3-4"
   amplifier   PM speakers
J1, J2 – RCA phono   T1 – audio transformer, pri.
      jacks   1000 ohms, sec. 8
                        ohms 250 mw.
Misc. - battery holder, enclosure, hardware, hookup
   wire, IC socket (14 DIP, low profile), knob, PC
   board, 2' shielded phono cable, 2 RCA phono
   plugs, two-conductor cable, etc.

Parts to Use: Optional Power Supply

BR1 – 1 a. 50 PIV bridge rectifier
C4 – 2000 mfd. 15 volt
P1 – 110 VAC plug
T2 – 6.3 volt 300 ma. transformer
Misc. – grommet, line cord, PC board, etc.

The PC Layout You'll Need

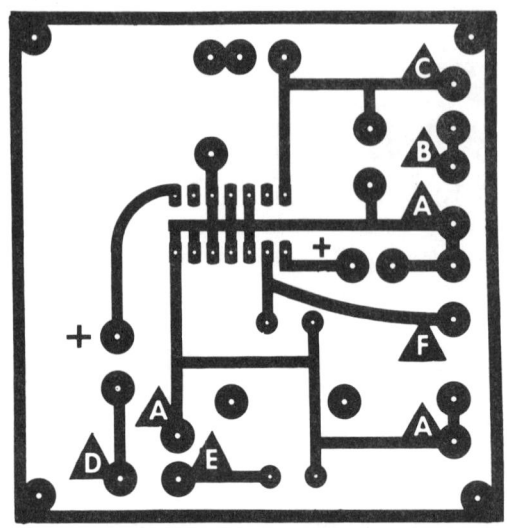

FIGURE 5—9B: INTERCOM SYSTEM PC LAYOUT (ACTUAL SIZE)

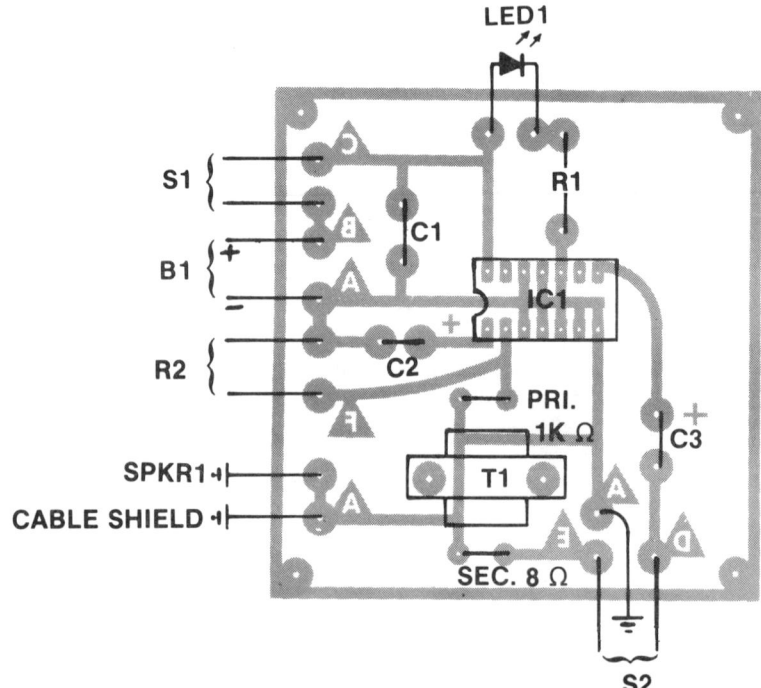

FIGURE 5—9C: INTERCOM SYSTEM COMPONENT PLACEMENT GUIDE (TOP VIEW)

Helpful Construction Hints

Insert and solder the component parts in the printed circuit board, carefully noting the polarities of C2 and C3. (See Figure 5—C.) Use of a low wattage soldering iron is recommended to avoid damage to the PC board. Since the IC pads are closely spaced, be sure that no bridges exist after soldering.

Shielded phono cable should be used to wire the following: S2 to J1, S2 to the primary of T1, and from pin 2 of the IC to R1. Separate the amplifier's input and output leads to minimize feedback and oscillations. When installing the IC in the socket, carefully note the pin basing configuration. A small dot is normally used to designate pin 1.

If the intercom is to be battery powered, six series connected pen light cells are preferable to a single nine volt battery, because of the high current requirements of the amplifier. An optional power supply may be incorporated into the master unit, eliminating the need for battery replacement. (See Figures 5—9D, 5—9E, 5—9F.)

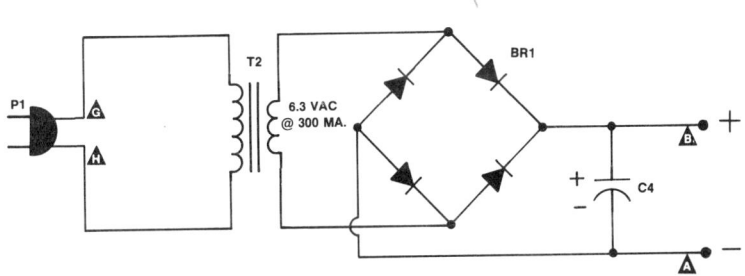

FIGURE 5—9D: INTERCOM SYSTEM OPTIONAL POWER SUPPLY

How It Works (Refer to Figure 5—9A)

Closing S1 supplies power to the IC amplifier module. S2 selects either speaker as an input or output to the amplifier. T1 serves as a matching transformer, coupling the low speaker impedance to the higher IC input impedance. R1 controls volume. When R1 is adjusted to minimum resistance, the output of T1 is bypassed to ground, reducing the signal voltage available to the amplifier input. C3, a DC blocking capacitor, couples the amplifier output to the speaker. Any spurious high frequency oscillation developed within the IC are suppressed by C1. C2, a bypass capacitor, improves amplifier gain.

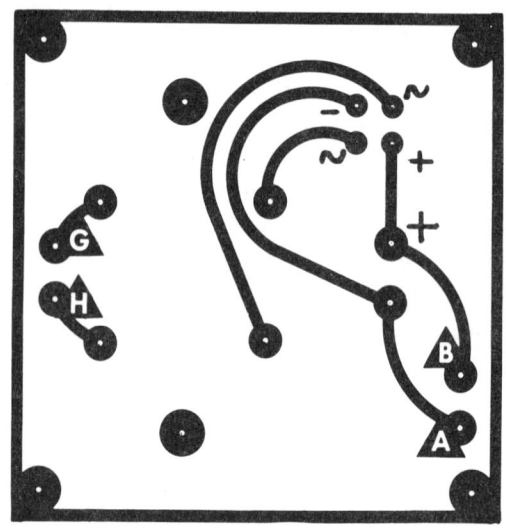

FIGURE 5—9E: INTERCOM SYSTEM OPTIONAL POWER
SUPPLY PC LAYOUT (ACTUAL SIZE)

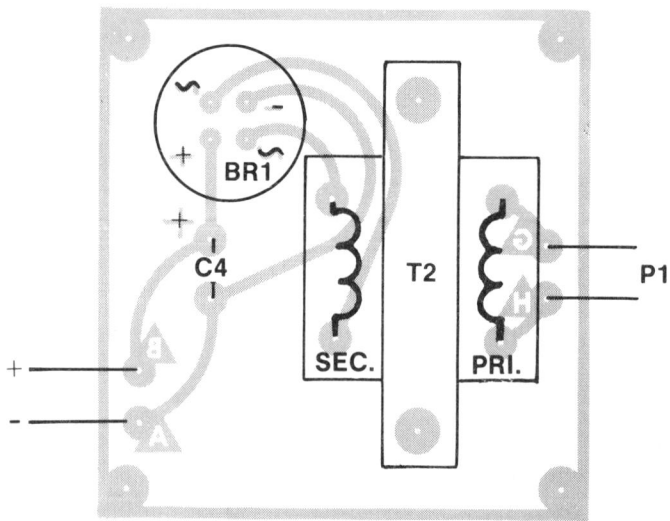

FIGURE 5—9F: INTERCOM SYSTEM OPTIONAL POWER
SUPPLY COMPONENT PLACEMENT
GUIDE (TOP VIEW)

Operating Tips

Mount the master and remote units in the desired locations. Interconnect the units using small two-conductor cable, with RCA phono plugs attached at each end. Separation up to several hundred feet is possible without an appreciable decrease in intercom performance. Avoid routing the cable near or parallel to any existing AC line wiring, which could induce hum and noise into the intercom system.

Close S1 to apply intercom power. Place S2 in the "talk" position. This allows the master unit to transmit to the remote speaker. Toggle S2 to the "listen" mode, and the remote now becomes the transmitter. Adjust R1 to achieve a comfortable listening volume. Open S1 when not using the intercom, to minimize battery consumption.

5—10: PLANT MOISTURE MONITOR

What It Does

If your houseplants are withering from thirst or are constantly soggy, the plant moisture monitor is the next best thing to having your own personal gardener. Never under- or overwater your greenery again! It has an easy-to-read meter, and you can tell at a glance when to water your plants. Build the plant moisture monitor, and have the lush greenery you've always wanted.

Here's the Schematic

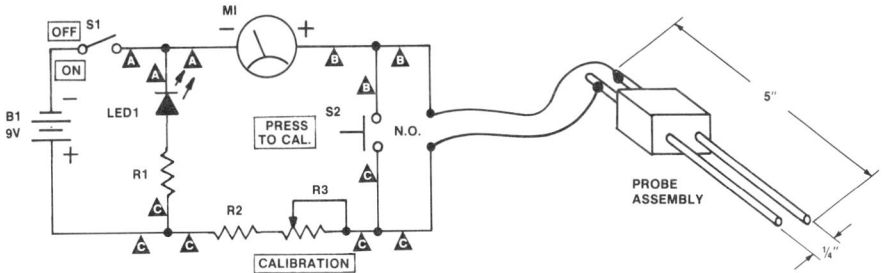

FIGURE 5—10A: PLANT MOISTURE MONITOR

Parts to Use

B1 – 9 volt transistor radio battery LED1 – jumbo red M1 – meter (0-1 ma, 1000 ohms) R1 – 220 ohm, ½ watt, 10% R2 – 6.8 K ohm, ½ watt, 10% R3 – 2.5 K ohm, linear taper S1 – SPST switch S2 – NO push button switch (spring loaded) Misc.- 9 volt battery connector, grommet, hardware, hookup wire, knob, probe (see text), terminal strips

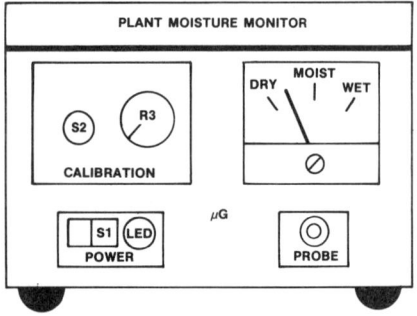

FIGURE 5—10B: PLANT MOISTURE MONITOR (FRONT PANEL VIEW)

Helpful Construction Hints

Mount the meter, switches, potentiometer, etc., in a minibox. (See Figure 5—10B.) Point-to-point wiring is recommended, with a terminal strip provided for R3 and the LED. Carefully observe the LED's polarity. The LED's cathode lead must be connected to the negative terminal of the battery and meter.

The sensing probe assembly is fabricated from two 5″ x ⅛″ metal rods, separated ¼″ by a wooden or plastic spacer. (See Figure 5—10A.) Small shishkabob skewers cut to length make excellent probes because of their corrosive-resistant plating and pre-sharpened ends.

Before attaching the lead wires to the sensing assembly, lightly tin the probe ends, using a soldering gun or a high wattage soldering iron. Avoid prolonged application of heat, which could damage the spacer block.

How It Works

When S1 is closed, power is applied to the LED and the series ohmmeter circuit consisting of R2, R3, M1, and the probe assembly. (See Figure 5—10A.) Momentarily closing S2 removes the probe's resistance, thereby permitting ohmmeter calibration by R3. Adjustment of R3 provides compensation for an "aging" battery and the resulting decrease in circuit current.

The probe senses relative soil resistance, which is dependent upon its moisture content. The more moist the soil, the lower its resistance. A lower resistance results in an increased circuit current as monitored by M1.

Operating Tips

Close S1. LED1 will light, indicating circuit power. Depress S2 and adjust the calibration control R3 for full-scale meter deflection (1 ma.). Release S2 and insert the sensing probe approximately 3″ into the soil being analyzed. The relative amount of moisture will be registered by M1. A small meter indication represents minimal moisture content; a large deflection, water rich soil. (See Figure 5—10C.) Replace B1 when a full-scale meter deflection can no longer be attained by adjustment of R2.

METER INDICATION	MOISTURE LEVEL
0 ma. - .2 ma.	extremely dry
.2 - .4	moderately dry
.4 - .6	moist
.6 - .8	very moist
.8 - .95	very wet

**FIGURE 5—10C: PLANT MOISTURE MONITOR
INTERPRETIVE GUIDE**

5—11: SLIDE PROJECTOR CONTROLLER

What It Does

If you would like to add a professional touch to your next home slide show, build the slide projector controller. By incorporating a low-cost IC timer, this controller automatically displays and advances your slides at preset intervals. Select a delay rate of between 10 and 90 seconds, or use the manual override feature. So, turn out the lights, sit back, and relax! Let the slide projector controller do all of the work while you enjoy the show.

Here's the Schematic

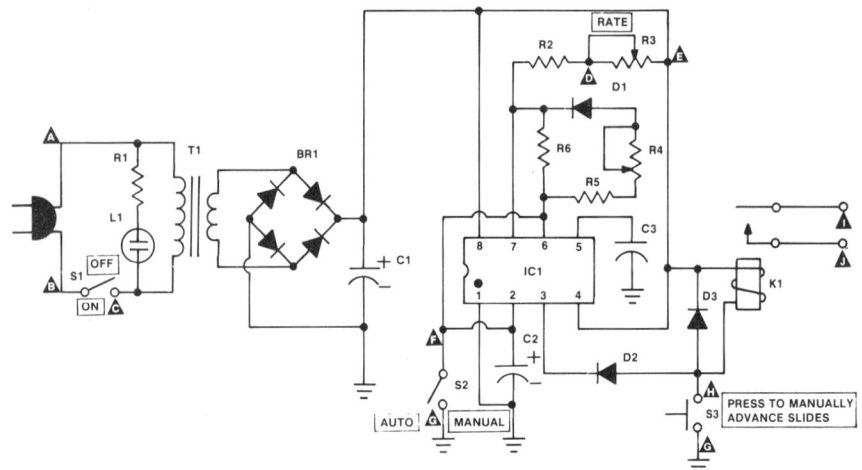

FIGURE 5—11A: SLIDE PROJECTOR CONTROLLER

Parts to Use

BR1 – 1a. 50 PIV bridge rectifier
C1 – 1000 mfd. 16 volt
C2 – 50 mfd. 12-15 volt tantalum
C3 – .01 mfd. disc. 25 volt
D1, D2, D3 – 1 a. 50 PIV
IC1 – 555 timer (Radio Shack 276-1723)

K1 – SPDT sensitive relay 6-9 VDC, 1 amp contacts @ 125 VAC (Radio Shack 275-004)

L1 – NE-2 neon light

P1 – 110 VAC plug

R1 – 56 K ohm, ½ watt, 10%

R2 – 150 K ohm, ½ watt, 10%

R3 – 2 meg ohm linear taper

R4 – 200 K ohm linear taper (PC mount)

R5 – 10 K ohm, ½ watt, 10%

R6 – 47 K ohm, ½ watt, 10%

S1,S2 - SPST switch

S3 – SPST NO push button switch (spring loaded)

T1 - 6.3 volt, 300 ma. transformer

Misc.– enclosure, hardware, hookup wire, IC socket (8 DIP low profile), knob, lamp lens, PC board etc.

The PC Layout You'll Need

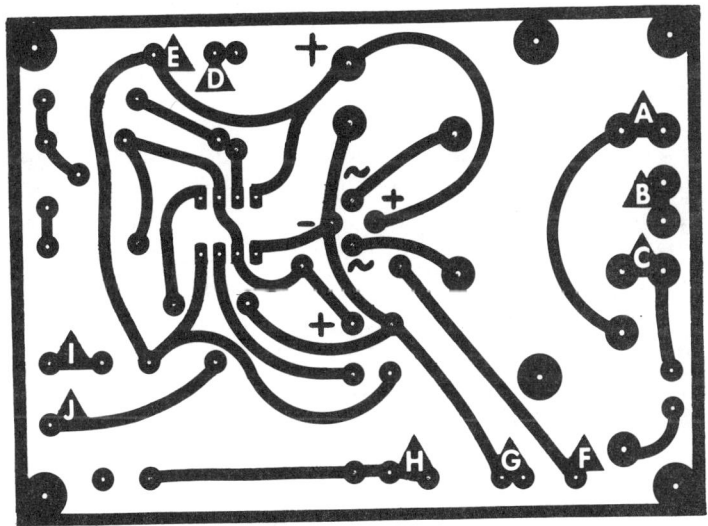

FIGURE 5—11B: SLIDE PROJECTOR CONTROLLER PC LAYOUT (ACTUAL SIZE)

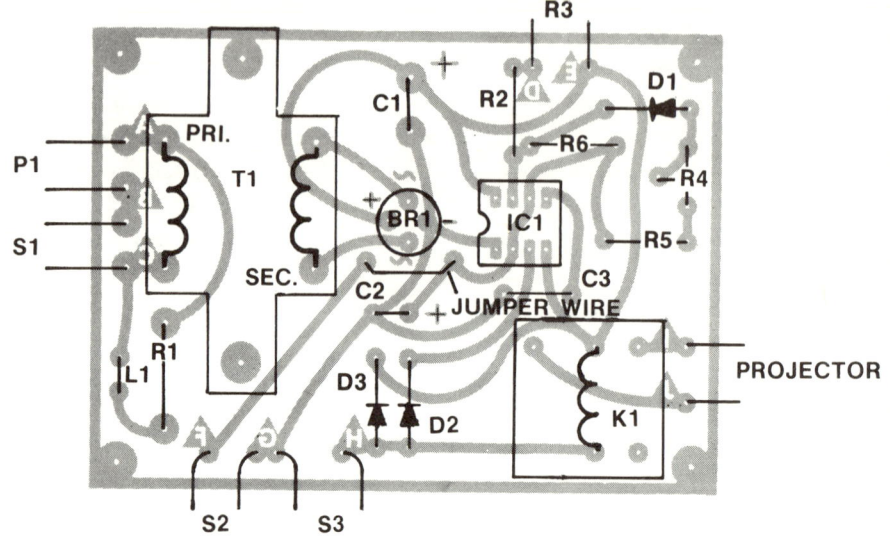

**FIGURE 5—11C: SLIDE PROJECTOR CONTROLLER
COMPONENT PLACEMENT GUIDE
(TOP VIEW)**

Helpful Construction Hints

Carefully observe the polarities of the capacitors and diodes when inserting the component parts in the printed circuit board. (See Figure 5—11C.)

Use of a low wattage soldering iron is recommended to prevent overheating and possible bridging of the closely spaced IC socket pads. Mount the switches, pilot light lens, rate control, and printed circuit board in a minibox. (Refer to Figure 5—11D.) Provide a small access hole in the enclosure to permit adjustment of the internally mounted control, R4. Be sure that the transformer's primary circuitry and relay contacts are electrically isolated from the enclosure. To prevent a possible shock hazard, verify this isolation with an ohmmeter before operating the controller. Using the high resistance scale, measure between the chassis and all points receiving line voltage. To prepare the unit for operation, connect the relay contacts across the existing remote control push button wiring. If the projector is not remotely controlled, install a small 110-volt solenoid within the unit to actuate the mechanism. (See Figure 5—11E.)

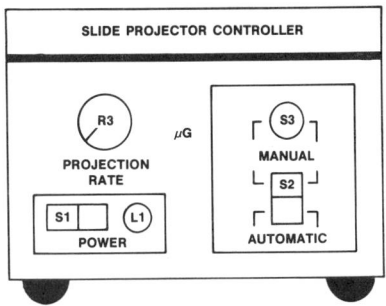

**FIGURE 5—11D: SLIDE PROJECTOR CONTROLLER
(FRONT PANEL VIEW)**

How It Works (See Figure 5—11A)

Closing S1 applies power to L1, and the full-wave power supply circuit consisting of T1, D1, D2, and C1. With S2 open, C2 will begin charging through the series resistors R2, R3, and R6. Initially, the trigger pin 2 is negative, causing the timer output pin 3 to be driven high (positive). Since the same potential exists across the relay coil's terminal, the relay will remain de-energized and its contacts open. When C1 charges to 2/3 of the applied voltage, circuitry within the timer will allow the capacitor to discharge through the parallel path consisting of R6, and D3, R4, R5. Simultaneously, during the discharge cycle, the timer output will be driven low (negative). This permits the relay contacts to close and actuate the projector. The contacts will again open when C1 has discharged to 1/3 of the power supply voltage. R2 controls the rate

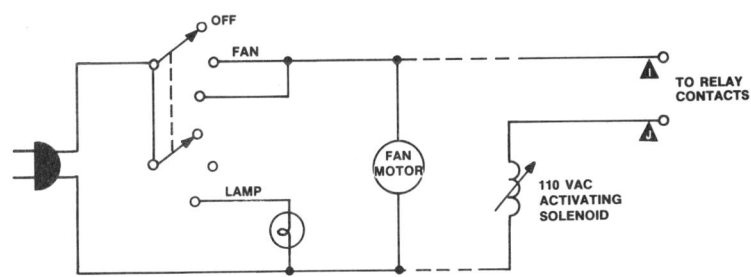

**FIGURE 5—11E: TYPICAL SLIDE PROJECTOR WIRING
DIAGRAM**

of slide selection. With R2 adjusted for maximum resistance, C1 will require the greatest charging time, resulting in the longest interval between cycles. Duration of projector actuation is accomplished by R4. When R4 is at minimum resistance, C1 rapidly discharges, and K1 is only briefly energized. S2 and S3 serve as operator "override" controls. When S2 is closed, C2 is held discharged, suspending timed operation. Manual slide selection is permitted by depressing S3.

Operating Tips

Close S1. L1 will be illuminated, indicating circuit power. Open S2 to provide automatic cycling. Adjust R2 and R4 for minimum resistance. With the projector operating, increase the resistance of the duration control (R4) until the projector receives a pulse of sufficient duration to advance the next slide. R4 should not require further adjustment, providing that the same projector is controlled. Adjust R2 to regulate the time each slide is displayed. Closing S2 and momentarily depressing S3 allows the projectionist to manually control the slide presentation.

6

Money-Saving Electronic Projects

6—1: NICAD REJUVENATOR-CHARGER

What It Does

Now you won't have to throw out those "worn out" expensive NiCd batteries any more. Pop them in the nicad rejuvenator-charger and before long the batteries will be like new again.

Each defective NiCd is rejuvenated by discharging a 5,000 MFD capacitor through the battery a number of times. Once the battery has been rejuvenated, it can be recharged to peak condition. Most batteries will respond to this treatment giving many more hours of reliable service.

Here's the Schematic

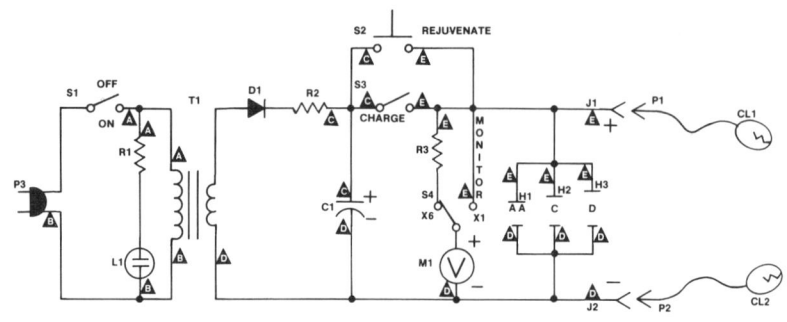

**FIGURE 6—1A: NICAD REJUVENATOR-CHARGER
SCHEMATIC**

Parts to Use

C1 – 5000 MFD, 30 V electrolytic	M1 – 0-5 DC voltmeter (500 Ω)
CL1 – battery clip, red insulated	P1 – banana plug, red
CL2 – battery clip, black insulated	P2 – banana plug, black
D1 – 2A 100 PIV silicon rectifier	P3 - AC plug
	R1 - 33K ½W, ±10%
	R2 - 350 ohm 5W, ±20%
H1 – AA (penlight) battery holder	R3 - 2500 ohm ½W, ±5%
H2 – C type battery holder	S1 - 1A SPST switch
H3 – D type battery holder	S2 - 3A DC NO pushbutton
J1 – banana jack, red	S3 - 3A DC SPST switch
J2 – banana jack, black	S4 - 1A SPDT spring return slide switch
L1 – NE-2 neon lamp	T1 – 117:24 VAC @ 1A transformer

Misc.– enclosure, grommets, hardware, lens cap and clip, strain relief, terminal strips, test lead wire, etc.

Helpful Construction Hints

Mount the parts on the front panel as shown in Figure 6—1B. Make all connections on the front panel parts' lugs and terminal strips. The wiring is straightforward with no special instructions.

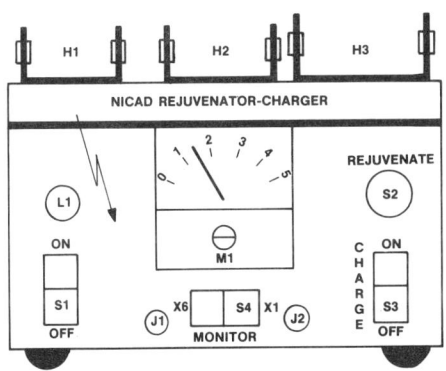

FIGURE 6—1B: NICAD REJUVENATOR-CHARGER FRONT PANEL LAYOUT

For convenience and ease of use, AA, C, and D size battery holders are fastened to the top of the assembly. Run the battery holder wires through grommets to the inside connections.

Make certain the X6-X1 monitor switch (S4) is a spring return switch so that it is always normally in the X6 position. This will protect the meter from excessive overload voltage if the rejuvenate or charge switches are activated with no battery load or if the battery is open.

How It Works

On-off switch (S1) applies 117 VAC to the power light (L1) and the step-down transformer T1. (Refer to Figure 6—1A.) The 24 VAC transformer secondary is rectified by D1 to DC voltage. Rejuvenator capacitor (C1) is a very large electrolytic, 5,000 MFD, that charges up to peak secondary voltage. During the rejuvenate mode C1 discharges through the NiCd battery under test blasting away any internal short. C1 also acts as a filter capacitor during the charge mode.

Once the short has been blasted out of the NiCd cell, the charge switch (S3) provides charging current. A good cell will charge to 1.25 V. The meter switch (S4) in the X1-monitor position reads NiCd potential. In the X6 position S4 converts the 5 V meter to a 30 V meter so that it can read voltage across an open output or an open NiCd cell.

The battery to be charged or rejuvenated-charged can be hooked to the output clips (CL1 and CL2), AA battery holder (H1), C battery holder (H2), or D battery holder (H3) depending on which is most appropriate.

Operating Tips

Most NiCd batteries are connected in series during normal operation. Also, they are usually kept in series while being recharged. The usual problem results because one of the batteries fails before the others. Since they are in series, one dead battery disables all of them. When this happens, the batteries are often tossed and replaced with a new set or, even worse, the appliance or whatever that is battery operated is discarded.

Here's how to restore the batteries to peak condition. Suppose a battery-operated grass trimmer won't accept a charge anymore. The clipper operates on two NiCd batteries. The first one is put in the nicad rejuvenator-charger and placed on "charge." Pressing monitor switch (S4) shows the output voltage at about ¾ volt and slowly but steadily increasing. Soon the cell voltage reaches 1.25 V. At this time turn off the charge switch while still watching the meter in the monitor position. If the voltage remains at 1.25 V the battery is probably okay and just needs further charging to be operational.

Now we'll place the second battery on the rejuvenator-charger. This time the meter reads zero volts. It looks like there's a shorted NiCd. Monitoring the meter for a few minutes while S3 is in the charge position confirms that the battery is not accepting a charge; the voltage remains at zero volts. It's time for rejuvenation. Open S3. Wait a minute and blast the NiCd with the rejuvenate switch. Wait a minute and repeat. Do this three or four times. Now flick S3 into the charge position while looking at the meter in the monitor position. If the voltage is rising towards 1.25 V you've probably repaired the battery. Continue testing and charging the battery as explained for the first battery. If the battery is still not accepting a charge try blasting it a few more times. You may still be successful.

Occasionally, a battery will charge to 1.25 V but then will slowly discharge with S2 and S3 open. If this happens, remove the battery from the rejuvenator and short circuit the negative and positive terminals together with a clip lead to discharge it. Put it back into the rejuvenator and blast again. This time it might work.

After a battery has been rejuvenated be sure to charge it for twelve or more hours before putting it back into normal service. Once restored, the former "dead" battery should give good service equal to the life of the other batteries that were used with it.

6—2: CAR BATTERY CHARGER-MONITOR

What It Does

Just plug this nifty project into your car cigarette lighter to charge your battery from the convenience of the passenger compartment. Or, with a flick of a switch you can easily monitor the car's electrical system.

Charge current is potentiometer controlled to a maximum of 2 amps. A built-in ammeter and voltmeter make it easy to keep an eye on both the charge and the monitor mode of this apparatus. Three LEDs serve as mode indicators and are labeled "ready," "charge," and "monitor."

Here's the Schematic

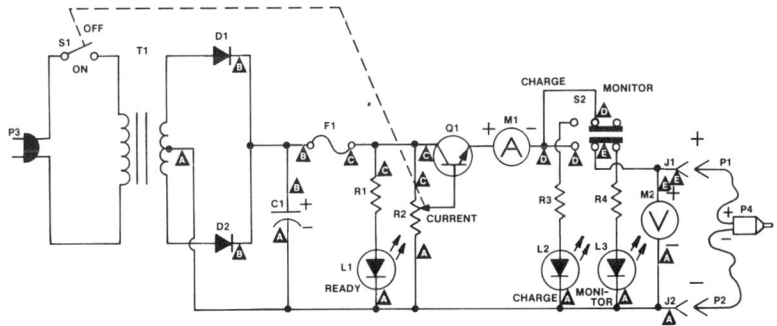

FIGURE 6—2A: CAR BATTERY CHARGER-MONITOR

Parts to Use

C1 – 1000 MFD, 50 V	P2 – banana plug, black
F1 – 3A fuse	P3 – AC plug
J1 – banana jack, black	P4 – cigarette lighter plug
J2 – banana jack, red	Q1 – MJE 3055 transistor
L1, L3 – LED, green	R1, R3, R4 – 1K 1W, ±10%
L2 – LED, red	R2 – 1K potentiometer, linear taper
M1 – 0-2A DC ammeter	S1 – SPST ganged with R2
M2 – 0-15V DC voltmeter	S2 – DPDT 2A DC
P1 – banana plug, red	T1 – 117:24 VAC, 2A
Misc. – enclosure, fuseholder, hardware, heatsink, knob, LED clips, silicon grease, strain relief, terminal strips, test lead wire, etc.	

Helpful Construction Hints

The front panel layout is shown in Figure 6—2B. Most of the parts can be mounted on the lugs of the front panel components. The rest of the connections can be made on a few strategically located terminal strips.

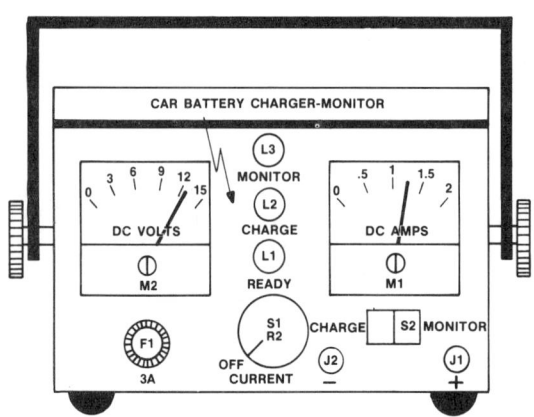

FIGURE 6—2B: CAR BATTERY CHARGER-MONITOR FRONT PANEL LAYOUT

Mount Q1 to a heat sink. The heat sink tab of the 3055 transistor is not electrically isolated from the collector so be sure to insulate the heat sink from the rest of the chassis. A little dab of silicon grease between the transistor heat sink tab and the heat sink will provide good thermal conductivity.

It's important to use a potentiometer ganged with the on-off switch. Hook the potentiometer so that output voltage is zero in the counterclockwise position. This will insure that the charger-monitor always starts out at zero when first turned on. You can then apply the optimum charge while watching the current meter.

Don't forget to hook up a pair of test leads to a car cigarette lighter plug; observe polarity. It makes it so handy to plug in to the cigarette lighter for making the connection to the car's electrical system.

You might want to reproduce Table 6—2A, gluing it to the outside case of the charger-monitor. These key voltages will provide helpful reference when monitoring your car's electrical system.

How It Works

Refer to Figure 6—2A for the following explanation. S1 applies power to the 117 V primary of T1. The secondary winding is stepped down to 24 VAC @ 2A. D1 and D2 provide full wave rectification converting the AC to pulsating DC. C1 acts as the filter capacitor to smooth out the DC. F1 protects the apparatus from accidental short circuit. Ready LED (L1) functions as a pilot lamp.

Current potentiometer (R2) controls base current to the transistor which in turn determines emitter-collector current (charge current). The ammeter (M1) measures charge current. In the charge position S2 connects the charger directly to the cigarette lighter plug via J1 and J2. The voltmeter (M2) reads charging voltage. The charge LED (L2) will also be activated.

When S2 is in the monitor mode the charger is disconnected from the output jacks. The charge LED will go out and the monitor LED will be lit. In this mode the voltmeter will monitor the car's voltage independent of the charger. In fact, the charger-monitor does not even have to be line powered. The monitor mode is convenient for checking battery voltage under various operating conditions as well as the car's charging system.

Operating Tips

It's a breeze to charge an auto battery. Simply plug the charger-monitor's cigarette lighter plug into the car's cigarette lighter receptacle. Rotate the on-off switch to the "on" position. Flip the charge-monitor switch to the "charge" position. Slowly turn the current pot clockwise while watching the ammeter. A one- to two-amp current will generally perk a battery up with an overnight charge. A note of caution: be sure to raise your hood so hydrogen gas does not accumulate in the engine compartment while charging.

Here's how to use the monitor mode. To check your car's electrical system plug in the charger-monitor to the car's cigarette lighter jack as previously explained for the charging mode. Switch S2 to the monitor position. Read the electrical system voltage from voltmeter M2. Table 6—2A shows typical voltages you would expect to read for a normal operating system.

Significant deviations from these typical voltages point to electrical system problems. For instance, if the battery voltage reads slightly low and dips considerably when the lights are turned on, the battery needs a charge or replacement. Excessive high or low voltage when the motor is running indicates voltage regulator or alternator problems. Refer to the author's *Complete Guide to Electrical and Electronic Repairs*, Parker Publishing Co. Inc., West

Electrical system power usage	DC Voltage (approx)	Comment
All devices off	12.5 - 12.8	Engine off Normal
Brake lights on	12.3 - 12.6	"
Parking lights on	12.1 - 12.4	"
Ignition on	11.7 - 12.3	"
Headlamps on	11.8 - 12.1	"
Starter on	9.5 - 10.5	Normal
Engine running	14.1 - 14.5	Charging voltage from alternator Normal

TABLE 6—2A: TYPICAL AUTO VOLTAGES

Nyack, N.Y. 10994, for a detailed explanation of automobile electrical system repairs.

6—3: RESISTOR/CAPACITOR SUBSTITUTION BOX

What It Does

How often have you rummaged through a parts box in search of a particular resistor or capacitor for experimenting or repair work? Wouldn't it be nice to have that much-needed part at your fingertips? You can, if you build the resistor/capacitor substitution box. Just a turn of a switch provides you with resistance of 22 ohms to 6.8 meg ohms and capacitor values from .001 mfd. to .47 mfd. Make your circuit design and troubleshooting tasks easier and more profitable by constructing this invaluable workshop project.

Here's the Schematic

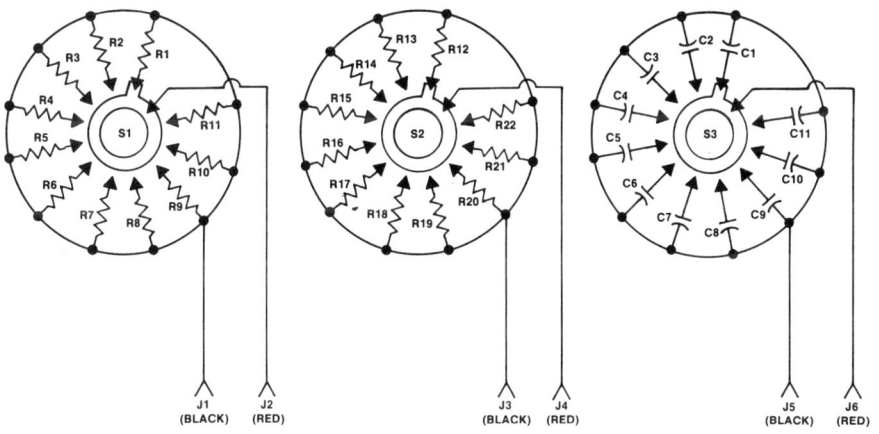

FIGURE 6—3: RESISTOR/CAPACITOR SUBSTITUTION BOX

Parts to Use

All capacitors in microfarads – 600 volts	
C1 – .001	C4 – .0068
C2 – .002	C5 – .01
C3 – .0047	C6 – .02

C7 – .047
C8 – .068
C9 – .1
C10 – .2
C11 – .47
J1, J3, J5 – 5-way binding posts (black)
J2, J4, J6 – 5-way binding posts (red)
R1 – R11 – values in ohms, 2 watt, 5%

R1 – 22
R2 – 47
R3 – 68
R4 – 100
R5 – 220
R6 – 470
R7 – 680
R8 – 1 K
R9 – 2.2 K
R10 – 4.7 K
R11 – 6.8 K
R12 – R 14 - values in ohms, 1 watt, 5%

R12 – 10 K
R13 – 22 K
R14 – 47 K
R15 – R22 - values in ohms, ½ watt, 5%

R15 – 100 K
R16 – 220 K
R17 – 470 K
R18 – 680 K
R19 – 1 meg
R20 – 2.2 meg
R21 – 4.7 meg
R22 – 6.8 meg
S1, S2, S3 – 1 pole, 11 position rotary switches
Misc.– enclosure, 1 foot 12 gauge copper wire, hardware,
        hookup wire, 3 knobs, etc.

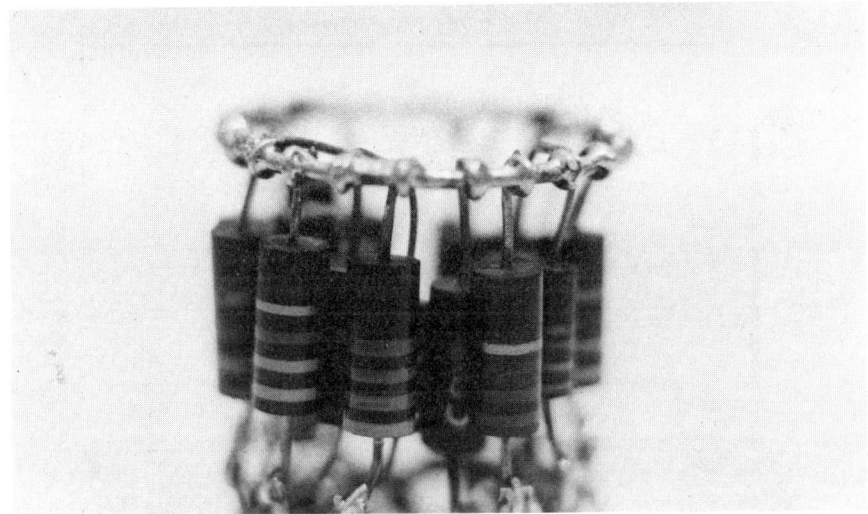

**FIGURE 6—3B: RESISTORS CONNECTED AND SOLDERED
TO COPPER BUSS RING**

Helpful Construction Hints

Rotate S1 fully clockwise (rear view). Beginning with the first switched position, mount and solder resistors R1 through R11. *Note*: Before soldering, be certain that the common lug is left open, since it will be connected to J2. Exercise care in attaching the resistor leads to the fragile switch terminals. Next, fabricate a buss ring from 12 gauge copper wire. This ring must be the same diameter as the switch terminals. After thoroughly cleaning the ring with steel wool, attach and solder each of the free resistor leads. (See Figure 6—3B.) To prevent possible resistor damage, heat sink each lead when soldering. Due to the rapid heat dissipation of the buss wire, use of a soldering gun is recommended. Follow the same outlined procedure for switches S2 and S3. Complete the construction phase by mounting each of the switch assemblies in the enclosure, and connecting the common switch terminals and buss rings to their respective jacks.

How It Works (Refer to Figure 6—3A)

S1 and S2 select the resistance available between J1, J2, and J3, J4 respectively. S1 provides resistor values rated at 2 watts from 22

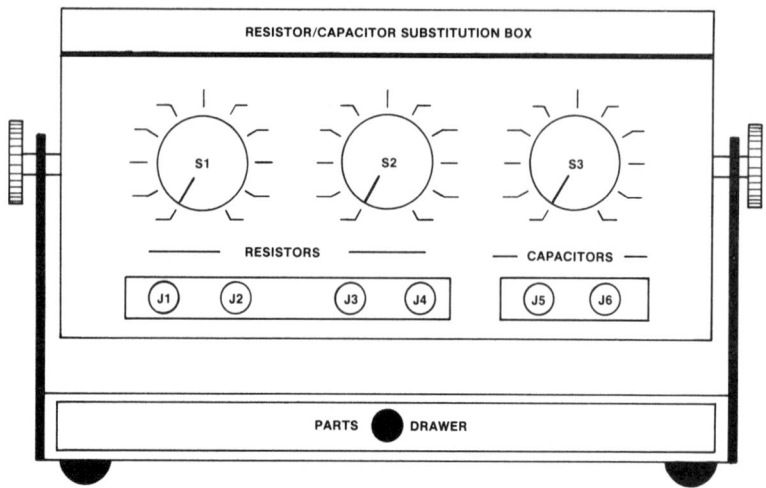

**FIGURE 6—3C: RESISTOR/CAPACITOR SUBSTITUTION
BOX (FRONT PANEL VIEW)**

ohms to 6.8 meg ohms. Values from 10 K ohms to 6.8 meg ohms are
selected by S2. Substitute capacitor selection is determined by the
position of S3—available at J5 and J6.

Operating Tips (See Figure 6—3C)

Capacitors: Use jacks J5 and J6. Select the desired capacitor
value with S3. *Note:* Be sure not to exceed the rated working voltage
of the capacitors.

Resistors: Select values of 22 ohms to 6.8 K ohms with S1,
utilizing J1 and J2. S2 provides resistance values of 10 K ohms to 6.8
meg ohms, available at J3 and J4.

To provide additional resistor values, connect the outputs of S1
and S2 in either series or parallel. By attaching a jumper wire
between J2 and J3, the resistance available between J1 and J4 will be
the total of the values selected by S1 and S2. For parallel operation,
connect jumper leads from J2 to J3 and J1 to J4.

Note: Make certain the resistor ratings are not exceeded,
particularly when using the lower ohmic values.

6—4: WALL OUTLET CHECKER

What It Does

This handy checker will allow you to quickly and conveniently analyze a three-wire wall outlet for safe and proper operation. Problems such as open neutral, open ground, and open hot wires, hot and neutral wires reversed, and hot and ground wires reversed are easily diagnosed by the checker's visual display. In addition, this outlet checker functions as a voltage tester, an invaluable aid in troubleshooting circuit problems.

Here's the Schematic

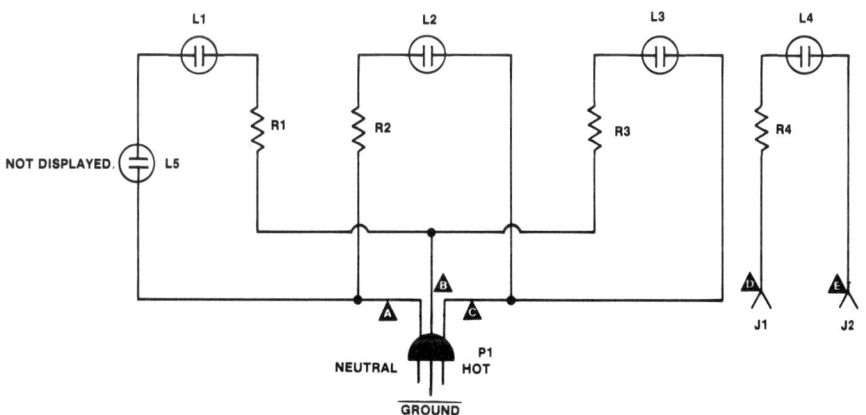

FIGURE 6—4A: WALL OUTLET CHECKER

Parts to Use

J1, J2 – phone tip jacks	R1 – 15 K ohm, ¼ watt, 10%
L1, L2, L3, L4, L5 –	R2, R3, R4, – 56 K ohm
NE-2 neon lights	¼ watt, 10%
P1 – 110 VAC plug	
Misc. - enclosure, hardware, hookup wire, insulating	
spaghetti, lenses, PC board, test probes, etc.	

The PC Layout You'll Need

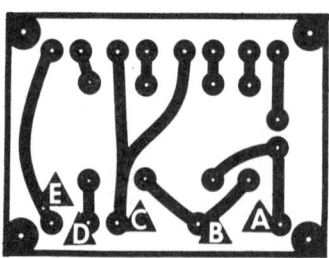

**FIGURE 6—4B: WALL OUTLET CHECKER
PC LAYOUT (ACTUAL SIZE)**

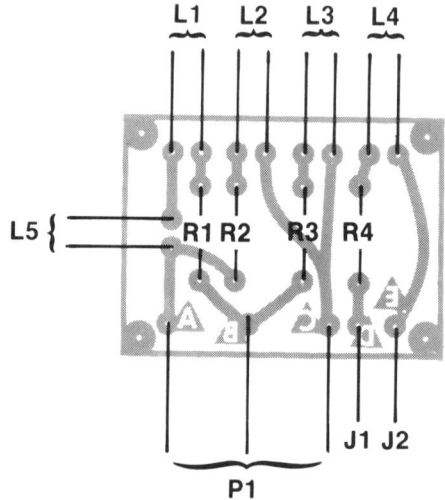

**FIGURE 6—4C: WALL OUTLET CHECKER COMPONENT
PLACEMENT GUIDE (TOP VIEW)**

Helpful Construction Hints

Mount all component parts in the printed circuit board, allowing full lead length for the display lamps L1, L2, L3, and L4 (see Figure 6—4C). The use of insulating spaghetti is recommended for the lamp leads. When positioning the display lamps in their respective lenses, avoid sharp bending of the lamp leads, particularly near the glass envelopes. *Note*: L5 improves circuit operation and should not be displayed.

P1 may be mounted in the enclosure or separately connected through a short length of three-conductor line cord. Since this is a line operated device, all circuitry must be electrically isolated from the enclosure.

How It Works (Refer to Figure 6—4A)

Correct circuit – A properly wired wall outlet provides a current flow through R2, lighting lamp L2. Since the ground and neutral wires are at the same electrical potential, a second path of conduction exists through R3, causing L3 to also light.

Open ground – Only L2 will light, with current limited by R2.

Open hot – Power is not available to light any of the display lamps.

Open neutral – L3 will be illuminated via current flow through R3.

Hot and ground reversed – L1 and L5 will light with current limited by R1. L3 will also be displayed by a current flow through R3.

Hot and neutral reversed – Power is applied to L1 and L5 through R1. A parallel path consisting of R2 and L2 is also created, causing L2 to glow.

Voltage tester – L4 will light when a voltage is applied to J1 and J2.

L1	L2	L3	CIRCUIT CONDITION
O	●	●	correct
O	●	O	open ground
O	O	O	open hot
O	O	●	open neutral
●	O	●	hot and ground reversed
●	●	O	hot and neutral reversed
light on ●			

FIGURE 6—4D: DISPLAY LIGHT INTERPRETATION CHART

Operating Tips

Plug the checker into the wall outlet being analyzed. A properly functioning outlet will be indicated by illumination of L2 and L3. Any other combination of lights will signal a malfunction and will require further inestigation. (Refer to Figure 6—4D.) Abnormal indications can be checked by physically examining the connections to the wall receptacle. Inspect for loose wires if an "open" indication exists. Check for proper wire placement if "wiring reversed" is indicated.

A white wire is used for the neutral line, and must be connected to the outlet's silver terminal. Black is commonly used for the hot wire, and is attached to the gold colored screw terminal. Green is reserved for the safety ground, and is connected to a corresponding green terminal on the receptacle.

If such checks fail to reveal the problem, the fault lies in the wiring or its connection in the fuse panel. This can be easily verified using the voltage tester function of the checker. With the checker unplugged, touch one test lead probe to the hot wire and the other to the neutral connection. L4 should light. Make the same test between the hot and ground wires, and L4 should light again.

6—5: ONE-HUNDRED-YEAR LAMP

What It Does

It makes light bulbs last and last and last. It's also very easy on energy. A nifty circuit converts an ordinary twin desk or table lamp (see Figure 6—5A) and ordinary light bulbs into a special bulb-electricity-saving device.

The one-hundred-year lamp has two switches mounted on its base. One switch is the normal on-off switch that is illuminated when the lamp is off so it's convenient to find the switch in a dark room. The other switch is a two-position switch. One position enables the lamp to act in the normal manner. The other position reduces energy consumption and increases bulb life many times over.

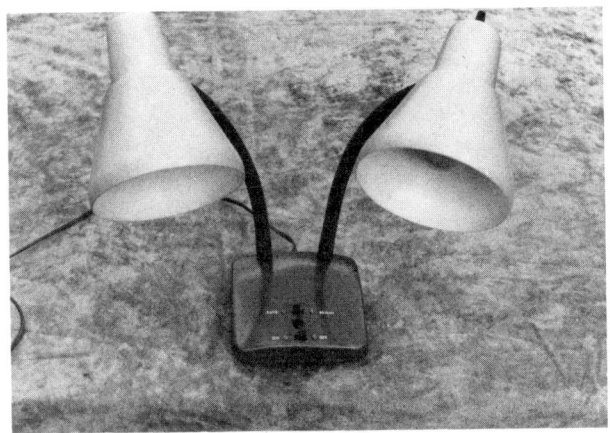

FIGURE 6—5A: ONE-HUNDRED-YEAR LAMP

Here's the Schematic

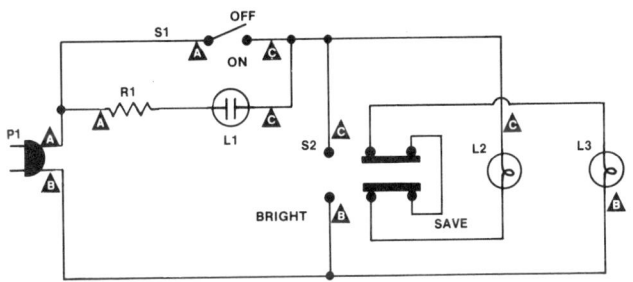

**FIGURE 6—5B: ONE-HUNDRED-YEAR LAMP
SCHEMATIC**

Parts to Use

L1 - NE-2 neon lamp	R1 - 56 K ½ W, ±10%
L2, L3 - 60W 120 V light bulb	S1 - SPST switch
P1 - AC plug	S2 - DPDT switch
Misc. - lamp sockets, line cord, neon lens cap and clip, strain relief, twin lamp assembly, etc.	

Helpful Construction Hints

It's easy to convert a standard twin lamp as shown in Figure 6—5A to a one-hundred-year lamp. Remove the bottom cover to mount both switches and the neon lamp in the top of the base. Position the neon lamp so that it illuminates the on-off switch. Label the other switch "bright" and "save" to identify both positions.

If you don't have a twin lamp around the house to modify, you can purchase an inexpensive one. Of course, you might want to build a lamp from scratch. Most discount houses stock all the lamp parts you will need. The sky's the limit on the design and style for your custom made lamp.

How It Works

Refer to Figure 6—5B for the following explanation. The resistor-lamp combination (R1 - L1) receives power when the switch is off. When switch S1 is turned on, R1 - L1 is shorted causing the neon to extinguish. When S2 is in the "bright" position, light bulbs L2 and L3 are connected in parallel resulting in normal operation. Flicking S2 to the "save" position changes the parallel lamp circuit to a series arrangement. Each bulb receives only one half of its normal voltage which is the key factor to long bulb life.

Operating Tips

There are many applications for a lamp such as this one. In the "save" position lamp brightness is reduced considerably. Some instances where lower brightness is advantageous are:

1. Night light
2. TV background illumination
3. Subdued room light
4. Bedroom lamp
5. House guard lamp
6. Nursery light

Remember that each time the lamp is operated in the "bright" position, there is normal wear and tear on the light bulbs. If the lamp is constantly used in the "save" position the bulbs should last just about indefinitely.

Use any wattage bulbs that you desire; be careful not to exceed the lamp maker's recommended wattage. The bulb wattage (brightness) is largely determined by your personal application for the one-hundred-year lamp.

6—6: TEST CENTER CONSOLE

What It Does

Here is a handy, multi-purpose project that no electronics workshop can afford to be without. If you are tired of the unsightly clutter of test speakers, antennas, etc. when testing or servicing radios, tape players, amplifiers, or CB radios, then build the test center console. This versatile unit provides you with a CB load and field strength meter, modulation monitor, speakers, and antenna in a convenient and easily portable package.

Here's the Schematic

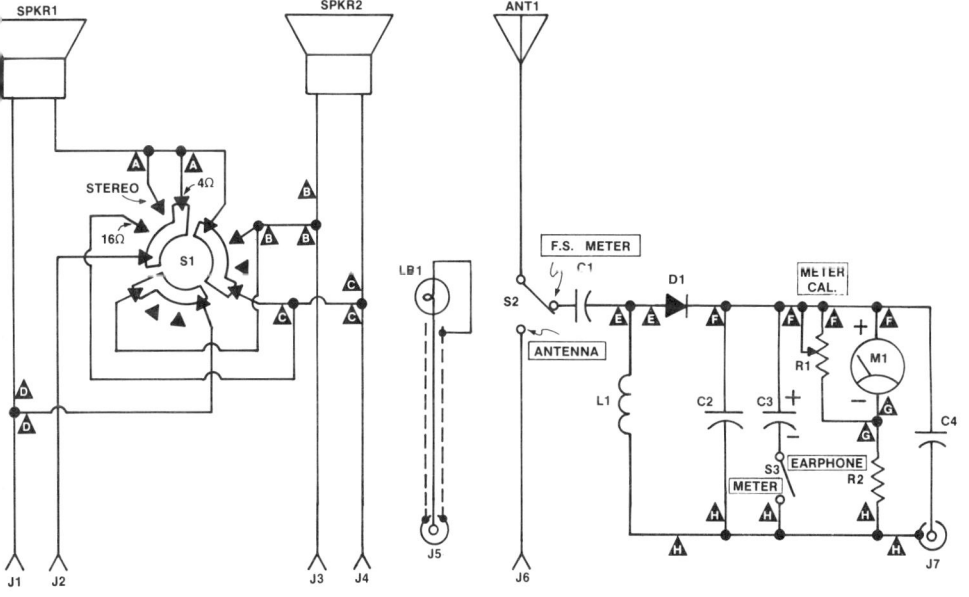

FIGURE 6—6A: TEST CENTER CONSOLE

Parts to Use

ANT1 – telescopic
    antenna
C1 – 470 mfd. 25 volt
C2 – .0015 mfd. 25 volt
C3 – 1 mfd. 6 volt
C4 – .22 mfd. 25 volt
D1 – 1N54 diode
J1, J3, J6 – 5-way binding
    post (red)
J2, J4 – 5-way binding post
    (black)
J5 – SO-239 coaxial
    connector
J7 – RCA phono jack
L1 – 2.5 mh. RF choke
LB1 – no. 47 light bulb

M1 – 0-1 ma. meter
    (1 K ohm)
R1 – 10 K ohm linear
    taper
R2 – 1 K ohm, ½ watt,
    10%
S1 – 3 pole, 3 position
    rotary switch
S2 – SPDT switch
S3 – SPST switch
SPKR1, SPKR2 – 3-4"
    8 ohm PM
    speaker

Misc. -bayonet socket, crystal earphone, enclosure, grommets, hardware, hookup wire, knob, terminal strips, 6" coaxial cable (RG-58/u), etc.

Helpful Construction Hints (See Figure 6—6B)

Install the speakers, switches, jacks, etc. in a suitable enclosure. Point-to-point wiring is recommended, with terminal strips provided for the field strength meter components. When wiring the field strength meter circuitry, avoid long lead lengths and be sure to observe the polarity of the diode, capacitor, and meter. A replacement telescopic antenna (radio, television) is suggested for ANT1. This antenna must be electrically isolated from the chassis. Provide a bayonet socket for LB1 and connect it to J5 through a short length of RG-58/u coaxial cable. To prevent damage to the cable's inner insulating sleeve, avoid overheating the braided conductors. Any high impedance crystal earphone may be used for checking modulation at J7.

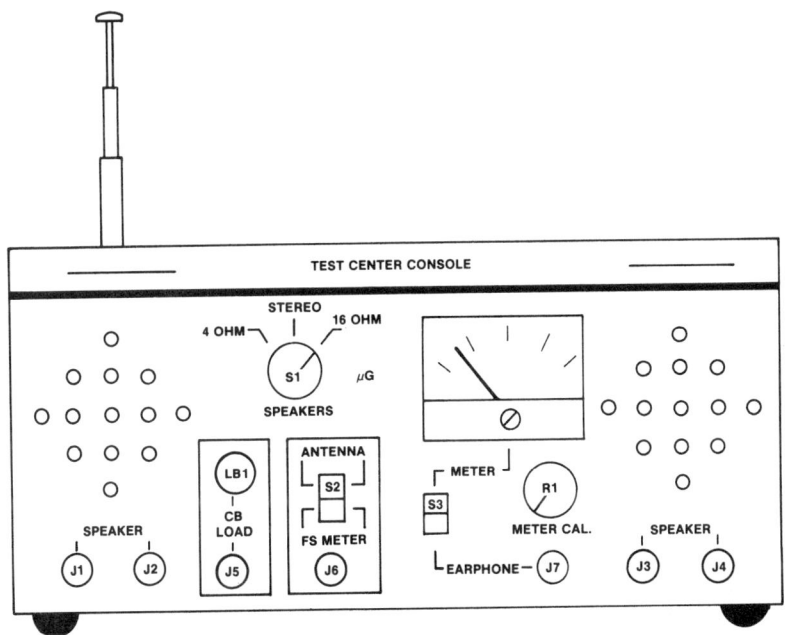

FIGURE 6—6B: TEST CENTER CONSOLE (FRONT PANEL VIEW)

How It Works (Refer to Figure 6—6A)

The test center console consists of four basic units—a multiple impedance speaker system, antenna, CB load, and field strength meter. S1 provides selection of two independent 8 ohm speakers or a single sound source of either 4 ohms or 16 ohms impedance. By switching SPKR1 and SPKR2 in parallel, an equivalent 4 ohm speaker is available at J1 and J2. Series connection of the speakers by S1 provides a speaker impedance of 16 ohms. To prevent possible damage to the CB under test, LB1 provides a 50 ohm antenna load. This light will vary in brightness relative to the amount of modulation. With S2 in the field strength position, a signal received by ANT1 is coupled to the meter circuit via C1. Component parts D1, C2, and C3 detect and filter the signal which is then applied to M1. Meter sensitivity is determined by R1, the calibration control. When R1 is adjusted for minimum resistance, all of the meter current is shunted through the control, resulting in no meter deflection.

Opening S3 allows the audio portion of the signal to be monitored by the earphone via C4.

Operating Tips

Substitute Speaker: Select the stereo position for S1 and use jacks J1, J2, and J3, J4 for the stereo operation. Utilize J1 and J2 for 4-Ohm or 8-Ohm speaker impedance as selected by S1.

Antenna: Connect the antenna input of a radio, receiver, etc. being tested to J6 and select the "antenna" mode for S2. Extend the telescopic antenna to obtain the best reception.

CB Load: Connect the antenna jack of a CB transceiver through a short length of coaxial cable to J5. When transmitting, LB1 will light and vary in brilliance relative to the amount of modulation.

Field Strength Meter: Close S3 and toggle S2 to the "F.S. Meter" mode. Extend and position ANT1 near the transmitting antenna. Adjust R1 for sufficient deflection of M1. This deflection will indicate the relative strength of the transmitted signal, and the amount of needle fluctuation will indicate the level of modulation. Using this technique, the CB can be adjusted for maximum performance. Opening S3 and connecting an earphone to J7 allows monitoring of signal quality.

6—7: HEADLIGHT MONITOR

What It Does

How many times have you left your car headlights on, only to return and find a dead battery? This will never happen again, when you build and install the headlight monitor. Turn your ignition off, and magically the monitor automatically extinguishes your headlamps. This easy-to-build circuit, incorporating a low-cost IC timer, delays shutoff (10-90 seconds), providing you with a safe and lighted exit from your car. Why wait? Eliminate the worry of a run-down battery and enjoy a convenience found only on today's luxury automobiles.

Here's the Schematic

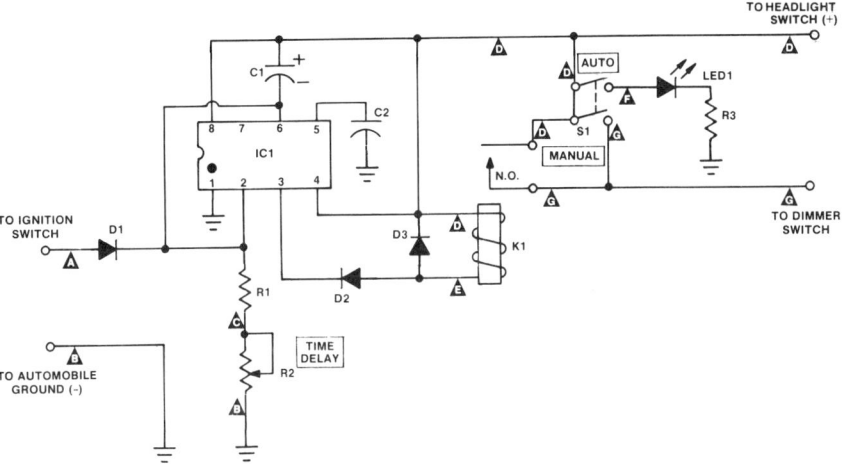

FIGURE 6—7A: HEADLIGHT MONITOR

Parts to Use

C1 – 47 mfd. 20 volt tantalum
C2 – .01 mfd. disc. 25 volt
D1, D2, D3 – 1 a. 50 PIV
IC1 – 555 timer
     (Radio Shack 276-1723)
K1 – 12 volt relay, DPDT, 10a. contacts
     (Radio Shack 275-218)
LED1 – jumbo red
R1 – 270 K ohm, ½ watt, 10%
R2 – 1 meg ohm linear taper
R3 – 470 ohm, ½ watt, 10%
S1 – DPDT 10 a. switch
Misc. – enclosure, hookup wire, IC socket (8 DIP low
     profile), knob, PC board, terminal strips, etc.

PC Layout You'll Need

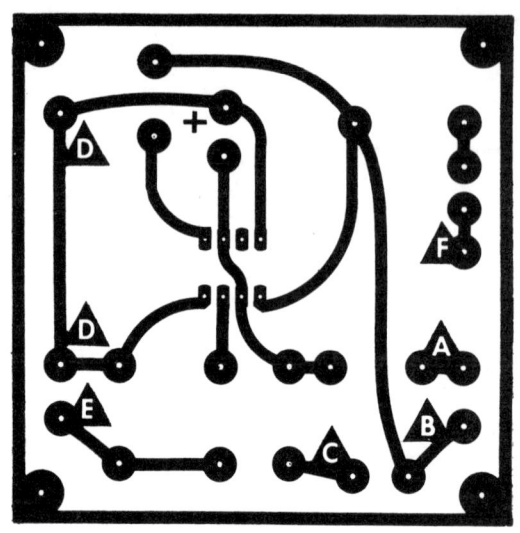

FIGURE 6—7B: HEADLIGHT MONITOR PC LAYOUT (ACTUAL SIZE)

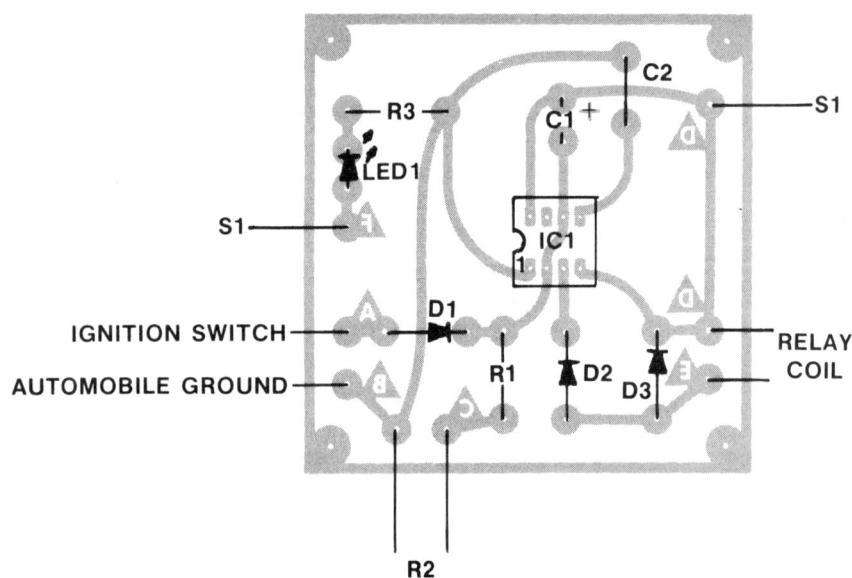

FIGURE 6—7C: HEADLIGHT MONITOR COMPONENT PLACEMENT GUIDE (TOP VIEW)

Helpful Construction Hints

Carefully solder the IC socket pads, using a low wattage iron. Avoid overheating and/or bridging the closely spaced clad. Insert and solder the remaining components, noting the polarities of C1, LED1, and the diodes. (See Figure 6—7C.) A tantalum type capacitor is recommended for C1. Aluminum capacitors tend to be "leaky" and consequently produce more erratic charging times. Because of the high current demands of the headlight circuit, it is suggested that the relay poles be parallel, to increase their current switching capability. When installing the IC in the socket, avoid bending or distorting the fragile leads. Also, pay particular attention to the basing configuration. A small dot is normally used to designate pin 1. Mount R2, S1, LED1, and the circuit board in a suitable enclosure. Provide an externally mounted terminal strip for connection to the automobile's electrical system.

How It Works (Refer to Figure 6—7A)

With S1 open and the headlight switch turned on, power is applied to the timing circuit. C1 is effectively discharged at this time and places a positive potential on the timer trigger, pin 2. The positive trigger drives the IC output low (ground potential), energizing the relay coil. The closed relay contacts now supply power to the headlight circuit. Simultaneously, C1 begins charging through the series resistance of R1 and R2. When the charge developed across C1 equals 2/3 of the battery voltage, the timer output will toggle high (positive), releasing the relay coil and turning off the lights. The value of C1 and the resistor network of R1 and R2 determine the timing duration. Adjusting R2 allows variation of the timed interval; minimum resistance—shortest time, maximum resistance—longest time. Placing the automobile's ignition switch in the "on" or "accessory" position maintains a positive timer trigger, and subsequently continues headlight operation. Turning the ignition switch off allows C1 to charge and initiate the timing cycle. S1 provides manual or automatic headlight operation. When S1 is closed, the automatic feature is defeated as indicated by illumination of LED1. D1, D2, D3, and C2 improve the monitor's

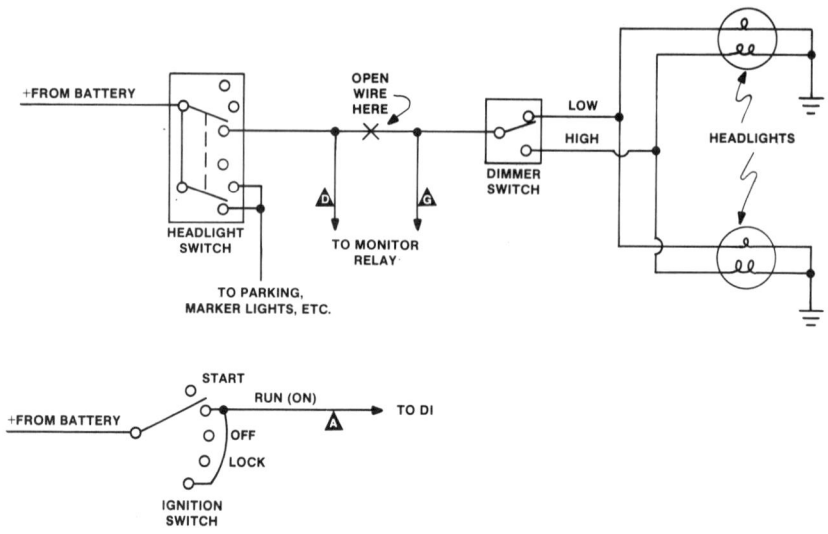

FIGURE 6—7D: TYPICAL AUTOMOBILE IGNITION AND LIGHTING CIRCUITS

reliability. D1 prevents a ground potential from reaching the trigger when the ignition switch is off. D2 and D3 suppress induced voltages from K1 that could cause a timer "latch-up." C2, used as a bypass filter, prevents false timer triggering.

Operating Tips

Installation: Disconnect the negative terminal from the automobile battery. Mount the headlight monitor in a convenient location under the dash panel. Be sure to allow easy access to the auto/manual and delay controls. For a "custom" installation, the monitor may be concealed in the instrument panel.

Next, locate the wire from the headlight switch to the floor mounted dimmer control. (See Figure 6—7D.) Open this lead and attach the switch and dimmer wires to Guide letters D and G respectively. Connect a wire from the ignition "on" terminal to Guide letter A. Verify with a voltmeter that power is not available at this point when the ignition is turned off. If access to the switch is difficult or "impossible," a connection to ignition controlled accessories (i.e. radio, heater) may be used. To complete the installation, provide an automobile ground connection to Guide letter B.

Operation: Open S1 (automatic mode). Turn on the ignition and headlight switches to provide continous headlight illumination. Return the ignition switch to the "off" position. The lights will remain on and then extinguish, with the time delay determined by the setting of R2. When S1 is closed, as indicated by the illumination of LED1, uninterrupted headlight power is provided. This feature is particularly useful when alignment is performed.

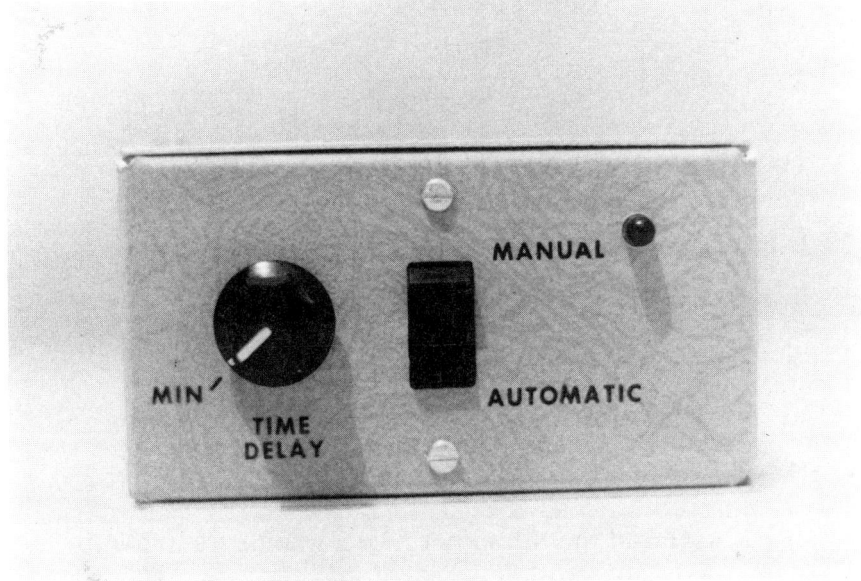

FIGURE 6—7E: HEADLIGHT MONITOR (FRONT PANEL VIEW)

6—8: DUAL POWER SWITCH

What It Does

Here is a simple money-saving project. The dual power switch will save you valuable energy by reducing the available power to lights and many appliances incorporating universal motors. Single speed hand power tools, such as drills, become more versatile when operated by the power switch. Save money by extending the life of incandescent lamps. Why not spend an evening building the dual power switch? No household can afford to be without this versatile, low-cost project.

Here's the Schematic

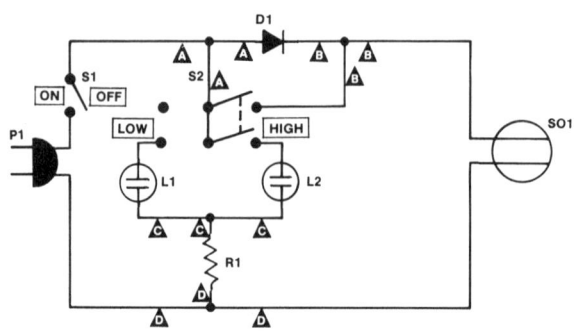

FIGURE 6—8A: DUAL POWER SWITCH

Parts to Use

D1 – 5 a. 200 PIV	S1 – SPST switch
L1, L2 – NE-2 neon lights	S2 – DPDT switch
P1 – 110 VAC plug	SO1 – 110 VAC outlet
R1 – 56 K ohm, ½ watt, 10%	
Misc.– assorted mounting hardware, enclosure, hookup wire, line cord, terminal strips, etc.	

Helpful Construction Hints

Mount the switches, outlet, lamp lenses, etc. in a minibox. Point-to-point wiring is recommended, with a terminal strip provided for the indicator lamps and their respective resistors. If a stud mount rectifier is used, furnish a suitable heat sink and be sure that it is electrically insulated from the enclosure. A safety grounded plug and outlet may be incorporated into the design, with a connection provided between the green colored terminals and the chassis.

Before operating the dual power switch, check all wiring to verify its electrical isolation from the minibox. Using the high resistance scale of an ohmmeter, measure between the enclosure and each ungrounded terminal connection. Infinite resistance should be registered at each point.

How It Works (Refer to Figure 6—8A)

Closing S1 provides one of two power levels, as selected by S2. With S2 in the "low" position, D1 will alternately conduct and impede the flow of current to any externally connected device. As the device is receiving only half of the available power, its output will decrease proportionately. Toggling S2 to the "high" mode effectively removes the rectifier from the circuit and permits full utilization of the line power. L1 and L2, with their respective current limiting resistor, serve to indicate the status of S2.

Operating Tips

Plug the device to be controlled into outlets S01, and provide 110 VAC line voltage for the power switch. Close S1, and select either low or high power with S2. L1 or L2 will light, indicating mode of operation.

Suggested applications include regulating incandescent lamp brightness, motor speed, and soldering iron temperature. Control of fluorescent lights and induction type motors is not recommended, since it is essential that they receive maximum power to function properly.

6—9: AUTOMATIC SHUT-OFF FOR BATTERY-OPERATED DEVICES

What It Does

Here is the perfect project if you are constantly replacing batteries in toys, games, or test instruments that were inadvertently left on. It is designed for any device using 6-15 volt batteries. The automatic shut-off for battery-operated devices provides power for 8-10 minutes, then turns them off. Why not eliminate costly and inconvenient battery replacement by building and installing this low-cost and easy-to-build project today?

Here's the Schematic

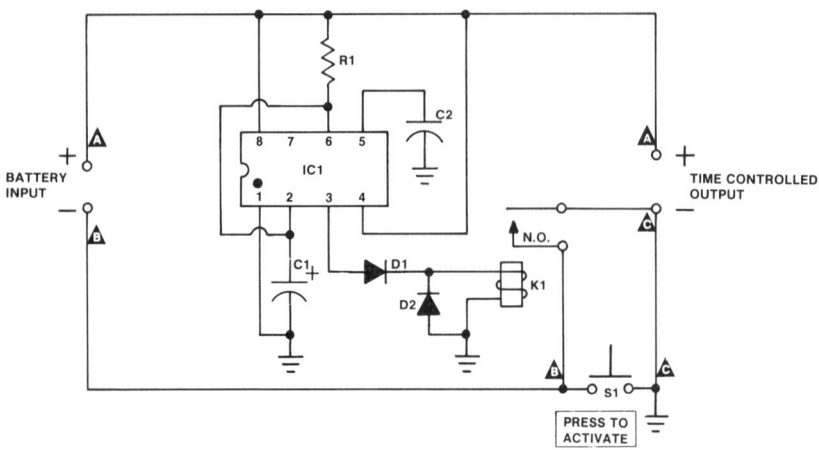

FIGURE 6—9A: AUTOMATIC SHUT-OFF FOR BATTERY-OPERATED DEVICES

Parts to Use

C1 – 100 mfd. 20 volt, tantalum
C2 – .01 mfd. 25 volt
D1, D2 – 1 amp, 50 PIV
IC1 – 555 timer

K1 – SPDT sensitive DC relay (Radio Shack 276-004)
R1 – 4.7 meg ohm, $\frac{1}{4}$ watt. 5%
S1 – SPST push button switch (spring loaded)

Misc. - hardware, hookup wire, IC socket (8 DIP low profile), PC board, etc.

The PC Layout You'll Need

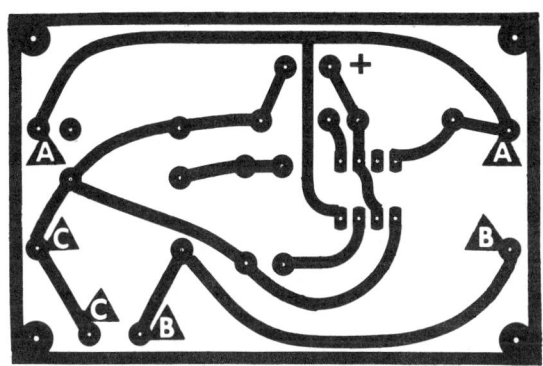

FIGURE 6—9B: AUTOMATIC SHUT-OFF FOR BATTERY-
OPERATED DEVICES (ACTUAL SIZE)

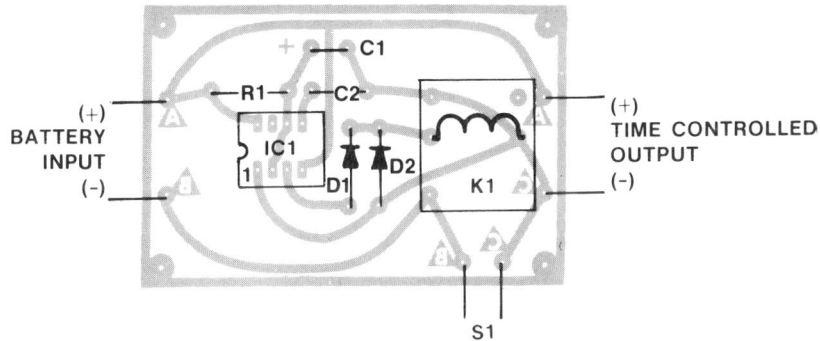

FIGURE 6—9C: AUTOMATIC SHUT-OFF FOR BATTERY-
OPERATED DEVICES COMPONENT PLACE-
MENT GUIDE (TOP VIEW)

Helpful Construction Hints (Refer to Figure 6—9C)

Insert and solder the component parts in the printed circuit
board. While soldering, avoid overheating and/or bridging the
closely spaced IC socket pads. When selecting C1, be sure to use a
tantalum capacitor, as aluminum electrolytic capacitors tend to be
"leaky," resulting in unpredictably longer timing cycles.

K1 can be any sensitive relay. Ideally, the holding current
should be small (2-12 ma.) to minimize battery consumption. If a

relay other than the suggested one is used, be certain that the normally open contacts are wired into the timing circuit.

How It Works (See Figure 6—9A)

Momentarily closing S1 supplies power to the timer and develops a negative pulse at the trigger, pin 2. This pulse causes the IC output to be driven high (positive), energizing the relay coil.

The closed relay contacts now maintain power to the timing and external circuit. C1 begins charging through R1. When 2/3 of the supply voltage is developed across C1, the timer will reset, allowing the output to return to ground potential. The relay is then released, removing battery voltage to the IC and the externally connected device. D1 and D2 prevent a circuit "latch-up," caused by induced voltages from K1. C2 provides bypass filtering, minimizing false timer triggering on positive pulses.

Operating Tips

Installation: Mount S1 and the timing circuit in the instrument or device to be controlled. For added installation flexibility, the relay may be remotely connected to the printed circuit board. Next, connect the device's battery leads to the input, and power leads to the output of the timing circuit. Be certain that the positive leads are attached to Guide letter A terminals on the printed circuit. In addition, the existing switch should be bypassed to prevent it from being inadvertently left open.

Approximate time* (in minutes)	C1	R1
1	47 mfd.	1 Meg Ohm
2-3	47 mfd.	3.3 Meg Ohm
5	100 mfd.	2.7 Meg Ohm
12	100 mfd.	6.8 Meg Ohm

*Note: The times given are approximate, due to variations in capacitor leakage, voltage rating, and component tolerances.

FIGURE 6—9D: TIMING DURATION CHART FOR VALUES OF C1 AND R1

Operation: Depress S1. Power to the device will be provided for approximately 8-10 minutes. Since the timing duration is determined by C1 and R1, their values may be changed to best suit the required "on" time of the device. (See Figure 6—9D.)

7

Projects
Just for
Fun

7—1: EMERGENCY SIREN

What It Does

Here's a loud penetrating electronic siren that's guaranteed to attract attention wherever it's heard. Protect your property with this neat 12 VDC siren. It's especially good for use as an automobile burglar alarm signaling device.

You'll find the construction easy. Three transistors, one IC, and a horn speaker comprise the major parts. A few capacitors plus a handful of resistors complete the parts requirement. Pop them in a PC board (Figure 7—1B) and you're ready to go.

Here's the Schematic

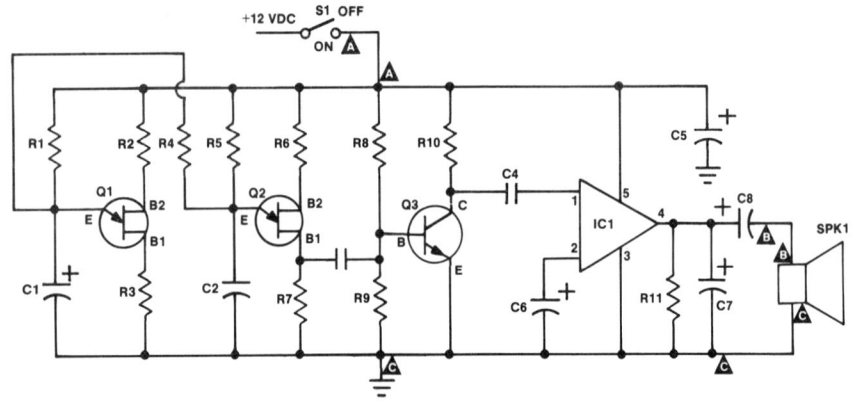

FIGURE 7—1A: EMERGENCY SIREN

Parts to Use

C1 – 10 MF, 25V
   electrolytic
C2 – .003 MF, 100 V
C3, C4 – .1 MF, 100 V
C5 – .5 MF, 16 V
   tantalum
C6, C8 – 100 MF, 25 V
   electrolytic
C7 – 2MF, 16 V
   tantalum
IC1 – LM383 audio
   amplifier
Q1, Q2 – ECG 6400
   unijunction
Q3 – EN2222 silicon
   transistor
Misc.– enclosure, hardware, heat sink, PC board, silicon
   grease, etc.

*NOTE: All resistors ½ or
   ¼ W

R1 – 33 K, ±10%
R2, R6 – 470 ohm, ±10%
R3, R7 – 47 ohm, ±10%
R4 – 470 K, ±10%
R5 – 220 K, ±10%
R8 – 27 K, ±10%
R9 – 3.3 K, ±10%
R10 – 1 K, ±10%
R11 – 220 ohms, ±10%

S1 – SPST switch
SPK1 – 8 ohm horn speaker

The PC Layout You'll Need

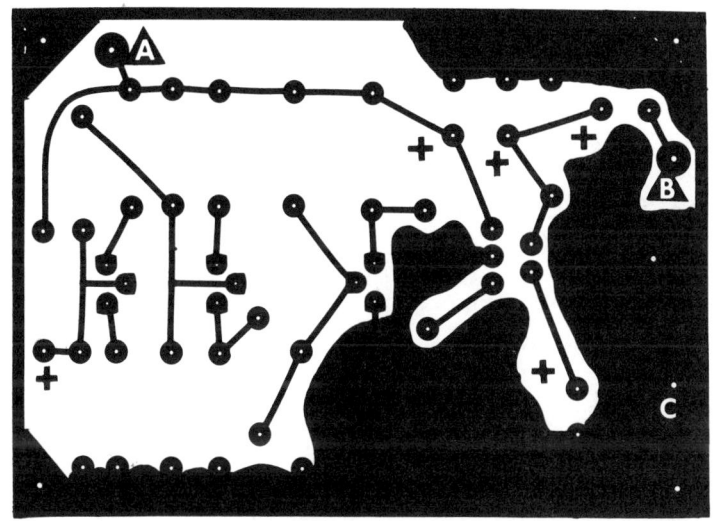

FIGURE 7—1B: EMERGENCY SIREN PC LAYOUT
(ACTUAL SIZE)

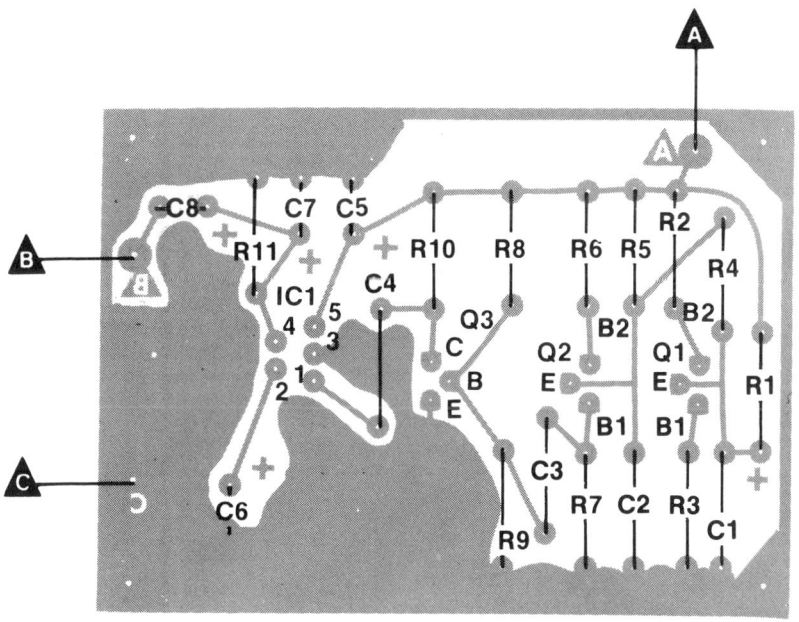

FIGURE 7—1C: EMERGENCY SIREN COMPONENT
PLACEMENT GUIDE (TOP VIEW)

Helpful Construction Hints

A small plastic box makes an ideal enclosure. Use a heat sink for the ground tab of IC1. A dab of silicon grease between the tab and the heat sink will help the heat transfer efficiently.

Although a conventional 8-ohm speaker works, you'll be more satisfied with the performance of a horn speaker. Many discount stores carry an inexpensive horn speaker made for indoor-outdoor PA use. A five-inch, five-watt type will really sound off. Mount the speaker so the horn is not muffled. For instance, if you are using it in a car, position the speaker inside the grill or other spot that has free air access to the outside.

How It Works

Please refer to Figure 7—1A for the following explanation. Close S1. Q1, and Q2 unijunction transistors form an oscillator circuit that generates an automatic rising and wailing signal. C1, R1, R4, R5 and C2 are the frequency determining components. The signal is then fed through coupling capacitor C3 to the base of the amplifier transistor Q3.

After amplification by Q3, the siren signal is coupled to pin 1 of the LM383 8-W audio amplifier IC through C4 for power amplification. Finally, the IC output (pin 4) is fed through C8 to drive the speaker.

Operating Tips

The siren draws around four to five hundred milliamps (.4 to .5A) of current while it is operating. This is a relatively low current draw for a siren so you won't have any power supply difficulty supplying it with a car battery. However, if you use this siren as part of an emergency warning system or burglar alarm operated on line voltage, be sure to use a 12 V power supply with enough amperage capacity to drive the siren circuits. Two series-connected 6 V lantern batteries will work nicely for portable operation.

7—2: STROBE LIGHT

What It Does

Here is a project that's just right for livening up your next home party. This variable rate strobe light, incorporating a high-intensity xenon flash tube, provides eerie slow-motion effects in a darkened room. In addition to the variable rate, a single flick of a switch offers you a choice of high or low brilliance. A few easy-to-make circuit changes convert this project into a manual strobe—ideal for photography.

Here's the Schematic

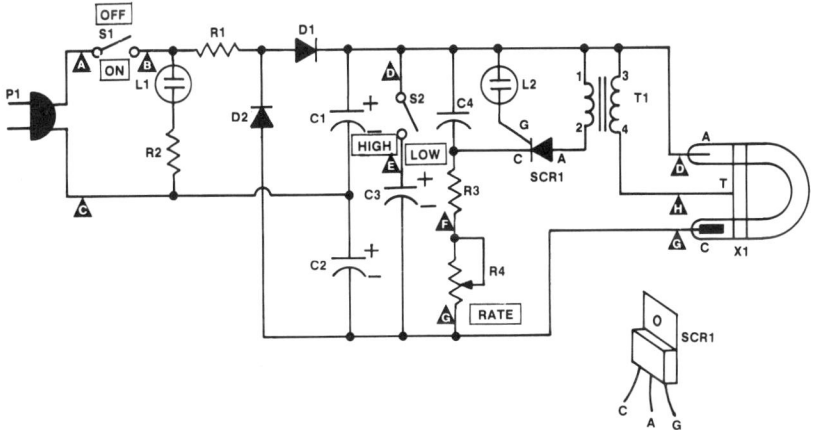

FIGURE 7—2A: STROBE LIGHT

Parts to Use

C1, C2 – 20 mfd, 150 volt	R3 – 330 K ohm, ½ watt, 10%
C3 – 10 mfd, 350 volt	R4 – 2 megohm linear taper
C4 – .47 mfd, 400 volt	S1, S2 – SPST switches
D1, D2 – 1 a. 200 PIV	SCR1 – 4 a. 200 PIV, .8 volt
L1, L2 – NE-2 neon lights	gate (GE C106B1)
P1 – 110 VAC plug	T1 – 4 KV trigger coil
R1 – 50 ohm, 5 watt, 20%	(Radio Shack 272-1146)
R2 – 56 K ohm, ½ watt, 10%	

X1 – xenon flash tube (Radio Shack 272-1145)
Misc. – enclosure, hardware, hookup wire, knob, line cord,
PC board, reflector (see text), terminal strip, etc.

The PC Layout You'll Need

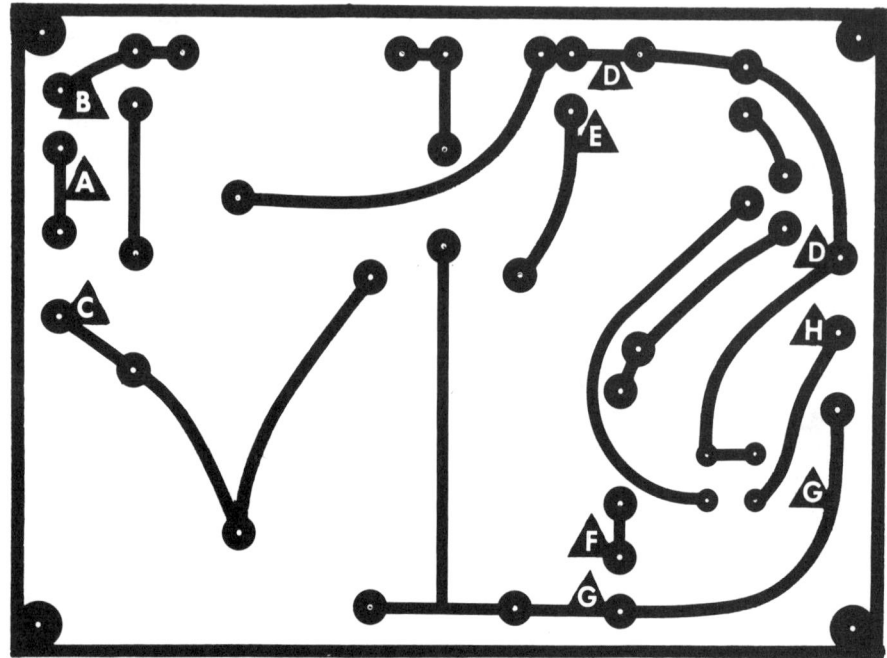

**FIGURE 7—2B: STROBE LIGHT PC LAYOUT
(ACTUAL SIZE)**

Helpful Construction Hints

Insert and solder the component parts in the printed circuit board, carefully noting the lead placement of the diodes, capacitors, and SCR. (See Figure 7—2C.) Proper orientation of the trigger coil (T1) also deserves special attention. The low resistance primary must be connected to the SCR's anode, and the higher resistance coil to the tube's trigger and anode (Guide letters H and D). (Refer to Figure 7—2A.) This can easily be determined using the low resistance scale of an ohmmeter.

Next, mount the switches, pilot light lens, potentiometer, and reflector in a suitable enclosure. A discarded photoflash or a lantern

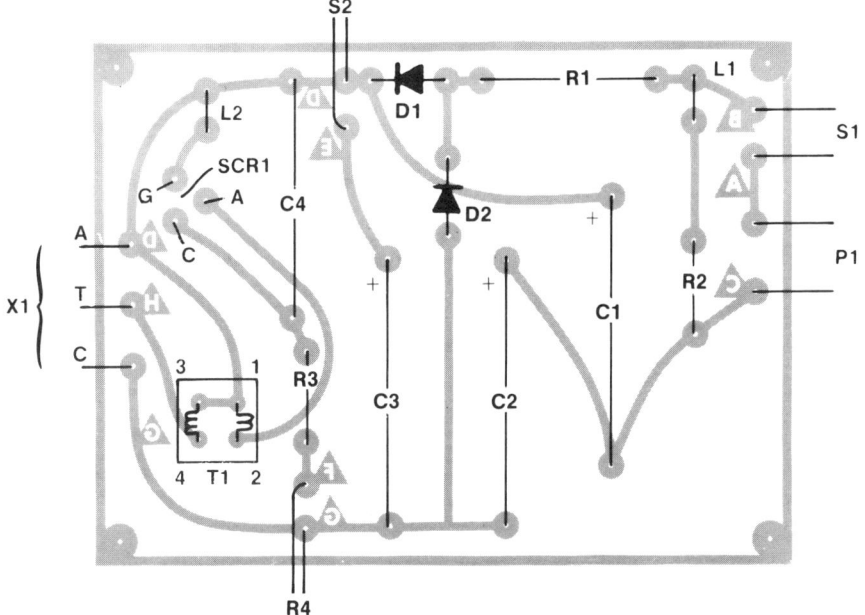

**FIGURE 7—2C: STROBE LIGHT COMPONENT
PLACEMENT GUIDE (TOP VIEW)**

reflector may be utilized. When installing the flash tube avoid bending the leads near the glass envelope. Due to the high pulse voltage (4 KV) of T1, the trigger lead should be separated or heavily insulated from the chassis, reflector, and any interconnecting wires. Finally, secure the completed printed circuit board, making sure that it is electrically insulated from the enclosure.

How It Works (See Figure 7—2A)

Closing S1 illuminates L1 with current limited by R2. Power is also applied to the voltage doubler circuit consisting of C1, D1, D2, and C2. C1 charges through D1 to the peak applied voltage. During the alternate half cycle, C2 will likewise charge via D2 and C1. Since C1 is effectively in series with C2, twice the peak input voltage will be developed across C2. Simultaneously, C4 will begin charging through R3 and R4. When C4 is sufficiently charged (50-60 volts), the SCR will conduct and permit C4 to discharge into the primary of T1. The stepped-up voltage at the secondary of T1 is now applied to the

FIGURE 7—2D: STROBE LIGHT

tube's trigger, causing the bulb to flash. The firing voltage of L2 (50-60 volts) determines the voltage required on C4 to bias the SCR into conduction. Closing S2 provides increased flash intensity because of the additional stored energy applied between cathode and anode of the flash tube.

Operating Tips

Caution—High Voltage. Avoid contact with the flash tube trigger circuit when testing strobe light operation.

Apply power with S2 open. Close S1 and adjust R3 for the desired flash rate. If a wider variance is desired, decrease and increase the values of R3 and R4 respectively. The rate should increase as the control is turned clockwise. If the converse is true, reverse the outermost leads of the rate control (R3). To increase the flash brilliance, close S2. However, prolonged high intensity operation will shorten the useful life of the flash tube, particularly at high flash rates.

To convert the strobe to manual operation, remove SCR1 from the circuit, and connect a jumper from the gate to the cathode pad. Then connect the manual switch leads to the anode and cathode points on the PC board. Each time the switch is closed, the strobe will flash.

7—3: COLOR ORGAN

What It Does

Here is a project no audio enthusiast should be without. When it is connected to your home sound system, the color organ transforms your music into a brilliant and dazzling light display. Treat your family and friends to a unique audio and visual experience, by building this popular project.

Here's the Schematic

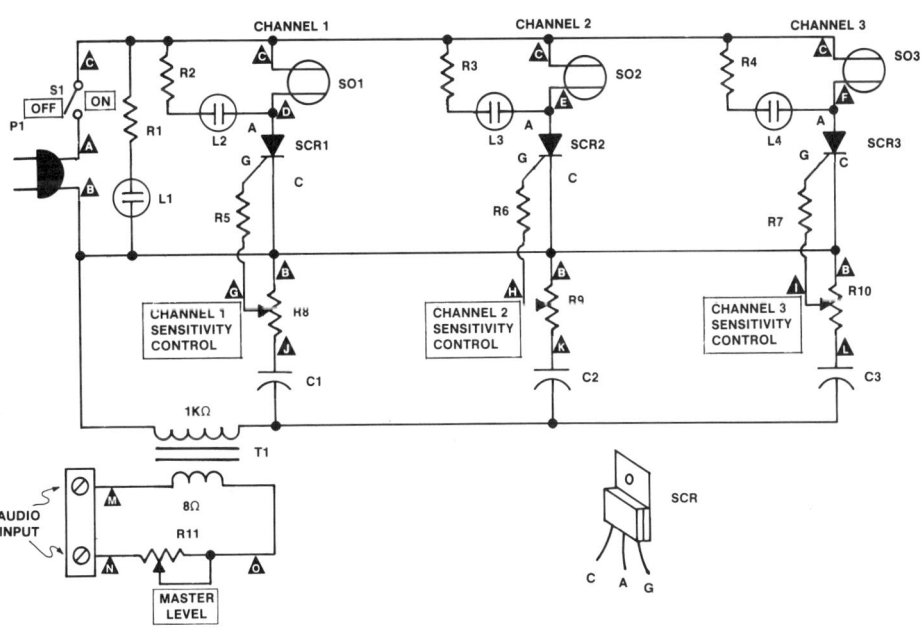

FIGURE 7—3A: COLOR ORGAN

Parts to Use

C1 – .22 mfd. 50 volt
C2 – .047 mfd. 50 volt
C3 – .022 mfd. 50 volt
L1, L2, L3, L4 – NE-2 neon lights
P1 – 110 VAC plug
R1, R2, R3, R4 – 56 K ohm, ½ watt, 10%
R5, R6, R7 – 330 ohm, ½ watt, 10%
R8, R9, R10 – 10 K ohm linear taper
R11 – 50 ohm linear taper
S1 – SPST switch
SCR1, SCR2, SCR3 – 4 a. 200 PIV, .8 volt gate
                  (GE C106B1)
SO1, SO2, SO3 – 110 VAC outlets
T1 – audio transformer, pri. 1000 ohms, sec. 8 ohms, AC 250
     mw.
Misc.– enclosure, grommet, hardware, knobs, lamp
     lenses, line cord, PC board, etc.

The PC Layout You'll Need

FIGURE 7—3B: COLOR ORGAN PC LAYOUT
                    (ACTUAL SIZE)

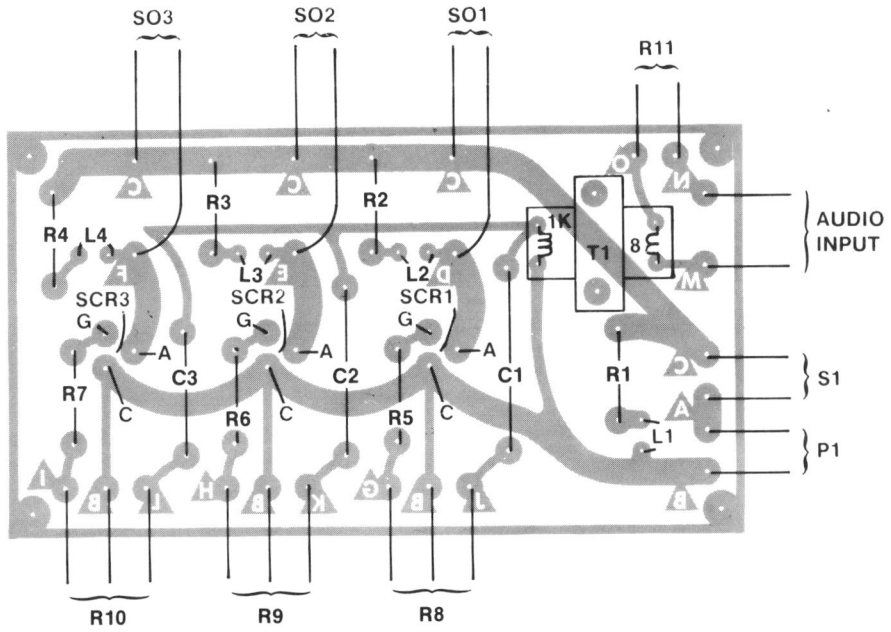

**FIGURE 7—3C: COLOR ORGAN COMPONENT
PLACEMENT GUIDE (TOP VIEW)**

Helpful Construction Hints

Insert and solder the component parts in the printed circuit board, carefully noting the lead placement of the SCRs. (Refer to Figure 7—3A.) To utilize the full power rating of each channel, a separate heat sink should be provided for each SCR anode tab. Application of silicon grease is recommended between the tab and heat sink to ensure maximum thermal conductivity.

Next, mount the outlets, controls, lamp lenses, and printed circuit in a medium sized enclosure. (See Figure 7—3D.) For convenience, the outlets may be installed in the rear apron of the chassis. When positioning the display lamps in their respective lenses, avoid bending the fragile leads near the glass envelope. Since this is a line operated device, be certain that all circuitry and the SCR heat sinks are electrically isolated from the enclosure.

How It Works (Refer to Figure 7—3A)

Closing S1 supplies power to L1 and properly biases the cathode-anode circuits of the SCRs on alternate half-cycles. To trigger the SCRs into conduction, a positive voltage must be applied to their gate leads. This positive gate voltage is obtained from the audio signal coupled through C1, C2, and C3. Since the capacitors are frequency selective, only certain audio frequencies will reach the gate terminals to turn on the SCRs.

The sensitivity controls, R8, R9, and R10, determine the gate voltage required for SCR triggering. R5, R6, R7 limit gate current. T1 provides impedance matching and electrical isolation of the amplifier and color organ circuitry. R11, the master control, regulates the amount of audio input applied to the primary of T1.

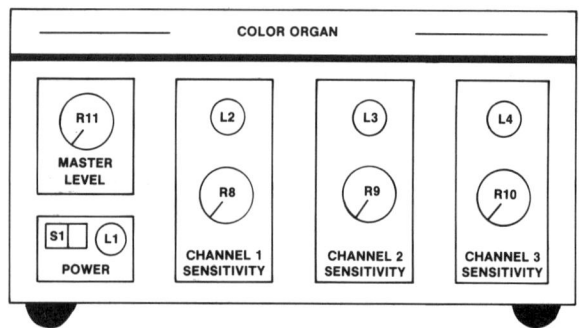

FIGURE 7—3D: COLOR ORGAN (FRONT PANEL VIEW)

Operating Tips

Connect any incandescent light source to each of the color organ's channels (maximum of 400 watts per channel). Small multicolored lights housed in a wooden enclosure with a light diffusing plastic front make a spectacular display.

Next, provide audio and power inputs to the color organ. Close S1 and adjust R3 for minimum resistance. While monitoring a record or tape recording, adjust the sensitivity controls R9, R10, and R11 respectively, until each display is independently modulated. Increase the resistance of R8 for decreased sensitivity of all channels at higher volume levels.

7—4: HAPPY CLOWN

What It Does

The happy clown is guaranteed to make a young child happy. It will keep him busy for hours. It's also a fine device for memory and color training. Kids love it!

A number of colorful lights, switches, and potentiometers are inserted into a clown's face (Figure 7—4A.) Eight switches control lights, buzzer, and a bell. Although the controls are installed in a random manner, children soon learn the exact function of each switch.

FIGURE 7—4A: HAPPY CLOWN PROJECT

If you want to be a hero to your son, daughter, nephew, niece, grandson, granddaughter, or the kid next door, this project is just the ticket. It's a toy that never will be forgotten.

Here's the Schematic

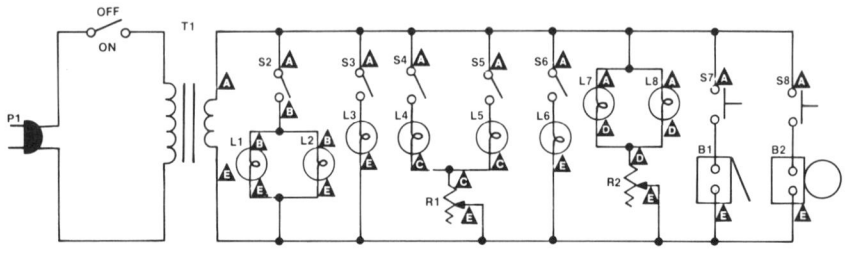

FIGURE 7—4B: HAPPY CLOWN SCHEMATIC

Parts to Use

B1 – 6 VAC buzzer	R1, R2 – 200 ohm, 10W WW potentiometer
B2 – 6 VAC bell	S1 – S6 – SPST switch
L1-L8 – #47 lamp	S7, S8 – NO momentary push-button switch
P1 – AC plug	T1 – 117:6.3 VAC, 2A transformer
Misc.– enclosure, hardware, knobs, line cord, pilot sockets and domes, strain relief, terminal strips, etc.	

Helpful Construction Hints

Figure 7—4C shows the component layout for the front panel clown's face. Use colorful lights and large handle switches for interesting and easy operation. Use lots of brightly-colored paint when making the clown's face.

A strong inexpensive enclosure can be made from ⅛″ hardboard. It's thin enough to mount all the front panel parts yet

**FIGURE 7—4C: HAPPY CLOWN'S FACE COMPONENT
LAYOUT**

rugged to take a lot of rough usage. It's also a good insulator so you
won't have to worry about any shock hazard.

Drill a pattern of holes close to the buzzer and bell so that their
sounds will not be muffled. Kids love to hear buzzer and bell
noises—parents don't. The hardboard tends to absorb the sound so
the happy clown should not be too unpleasant for the rest of the
family to hear.

How It Works

Refer to Figure 7—4B for the following explanation. On-off
switch (S1) applies power to the primary of T1, a 117 to 6.3 VAC
step-down transformer. The transformer secondary supplies 6.3
VAC to all the parallel loads.

Switch S2 controls the parallel ear lights (L1, L2). The nose
lamp (L3) is switched on and off by S3. The left eye is switched in by
S4 and the right eye by S5. The circuit of both eyes is completed
through potentiometer R1. Switch S6 controls the middle mouth
lamp. The two outside mouth lamps are potentiometer controlled
by R2. Buzzer (B1) and bell (B2) are activated by push button
switches S7 and S8 respectively.

Operating Tips

Introduce the happy clown to the child and the rest will be automatic. The bottom row of switches is arranged in a random order for added interest. The right hand bottom switch is the master on-off switch causing all controls to be inoperable until it is thrown. It won't take long for the child to learn the switching combinations and operate the clown like a pro.

Once the child learns the switch and light combinations, educational games can be invented and played. For instance, games involving colors, sequences, multiple orders, etc. can be played on the happy clown. Kids have great fun and soon will be using the happy clown for all kinds of make-believe things from a fire engine to a space command control panel.

7—5: WINKING NIGHT LIGHT

What It Does

The winking night light, Figure 7—5A, is a faithful night time companion. It doesn't do much but sit on a dresser and merrily blink away. Kids love it. In fact, if you are building one of these little jewels for one of your kids be sure to make one for each child—or you'll have a battle on your hands.

FIGURE 7—5A: WINKING NIGHT LIGHT

Before jumping into bed for the night, flick on the winking night light switch. Four neon lamps will alternately flash on and off giving a subtle but unique light display. Surprise someone with a gift of a winking night light. It's guaranteed to please.

Here's the Schematic

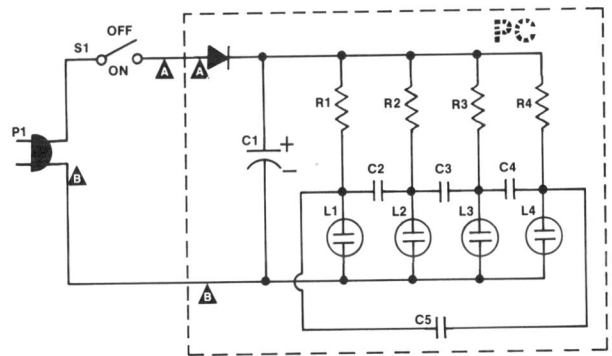

FIGURE 7—5B: WINKING NIGHT LIGHT SCHEMATIC

Parts to Use

C1 – 25 MF, 150 V electrolytic	L1, L2, L3, L4 – NE-2 neon lamp
C2, C3, C4, C5 – .47 MF, 200 V	P1 – AC plug
D1 – .5A, 400 PIV silicon rectifier	R1, R2, R3, R4 – 1.8 meg ½W, ±10%
	S1 – SPST switch

Misc. – enclosure, hardware, line cord, PC board, plastic front cover-diffused, etc.

The PC Layout You'll Need

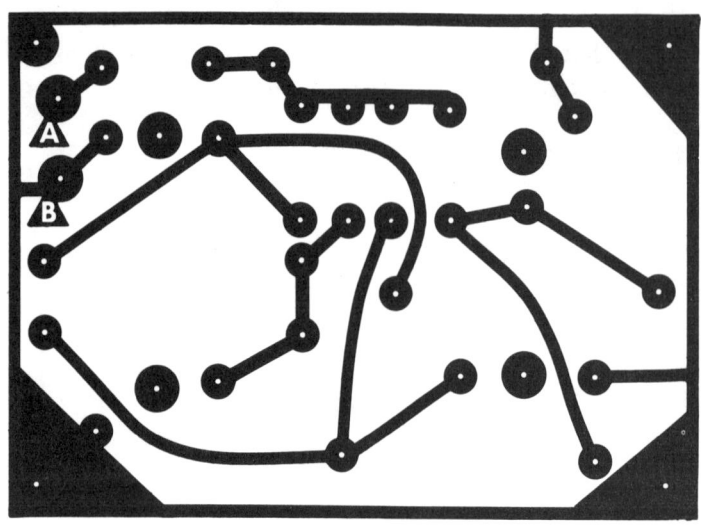

FIGURE 7—5C: WINKING NIGHT LIGHT PC PATTERN
(ACTUAL SIZE)

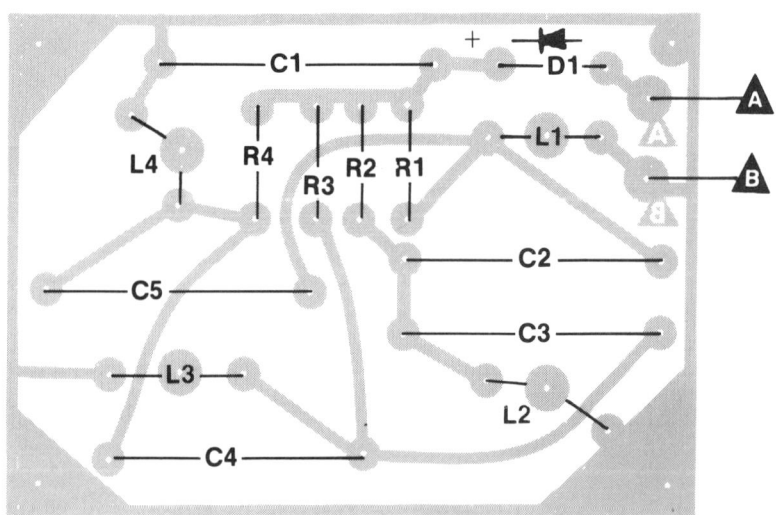

FIGURE 7—5D: WINKING NIGHT LIGHT COMPONENT
PLACEMENT GUIDE (TOP VIEW)

Helpful Construction Hints

A 3″ × 4″ bakelite plastic box makes an ideal enclosure. Mount the on-off switch in the side or back of the box. All the other parts are connected to the PC board.

Fasten the four neon lamps so that they stick out of the four holes in the foil side of the PC board. (Refer to Figure 7—5D.) Mount the completed PC board part side down into the plastic box. Cut a piece of clear or colored diffused plastic to act as the box cover and front panel.

How It Works

Refer to Figure 7—5B for the following explanation. On-off switch (S1) applies AC power to half wave rectifier (D1) which changes the AC to pulsating DC. Filter capacitor (C1) smoothes out the pulsating DC.

One of the neon lamps will light causing the capacitor connected to it to charge. As the capacitor charges it reaches the firing voltage of the next neon, about 47 V. The next neon then fires, discharging the capacitor, reducing the voltage across the first neon so that it extinguishes. As soon as the second neon ignites, the next capacitor will begin charging up through it until the third neon fires. The lighting of the third neon turns off the second. The cycle repeats for the last neon and returns to the first neon to start the whole process over again.

Operating Tips

The full effect of the neon light action occurs only in a completely dark room. Don't worry if the winking night light is left on inadvertently during the day time. It uses a minute amount of power and will have very little, if any, effect on your electric bill. The neon lamps are extremely reliable and should last for many, many years.

7—6: ELECTRONIC CRYSTAL BALL

What It Does

Here is the electronic update of the age-old coin toss. Add sophistication to your decision-making process with the electronic crystal ball. Simply press a button and this project randomly selects one of two choices with LED indicators. Supplemented by optional sound accompaniment, the electronic crystal ball can be used for sporting events, games, or just for fun.

Here's the Schematic

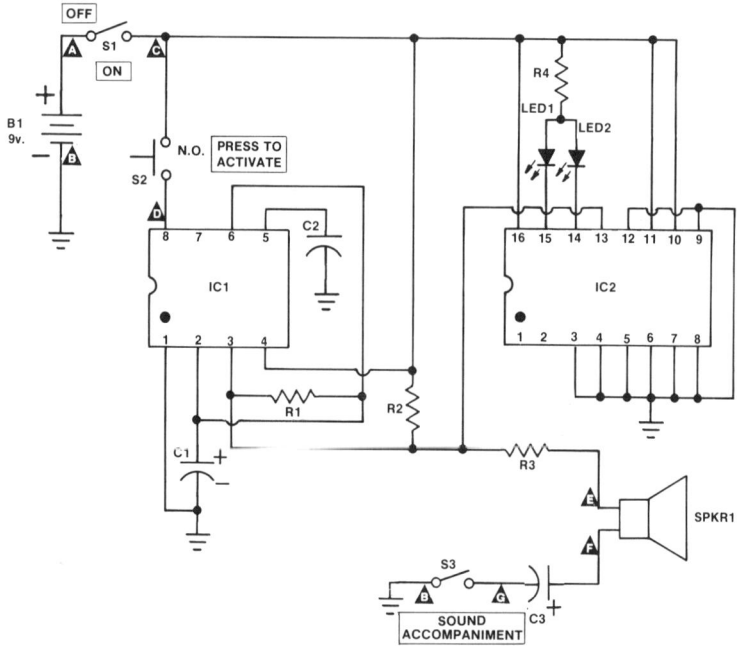

FIGURE 7—6A: ELECTRONIC CRYSTAL BALL

Parts to Use

B1 – 9 volt transistor
    radio battery
C1 - 1 mfd. 15 volt
    electrolytic
C2 – .01 mfd. 25 volt
C3 – 10 mfd. 15 volt
    electrolytic
IC1 – 555 timer
IC2 – 4027 J-K flip-flop
LED1 – miniature red
LED2 – miniature green

R1 – 6.8 K ohm, ½ watt, 10%
R2 – 1 K ohm, ½ watt, 10%
R3 – 47 ohm, ½ watt, 10%
R4 – 220 ohm, ½ watt, 10%
S1, S3 – SPST switch
S2 – SPST push button switch
    (NO spring loaded)
SPKR1 – 8 ohm, 2-4",
    PM speaker

Misc. - battery clip, enclosure, 8 and 16 DIP IC sockets
    (low profile), hookup wire, mounting hardware,
    PC board, etc.

The PC Layout You'll Need

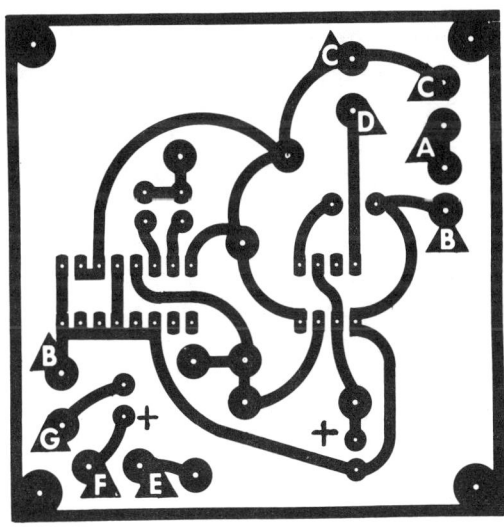

FIGURE 7—6B: ELECTRONIC CRYSTAL BALL PC
    LAYOUT (ACTUAL SIZE)

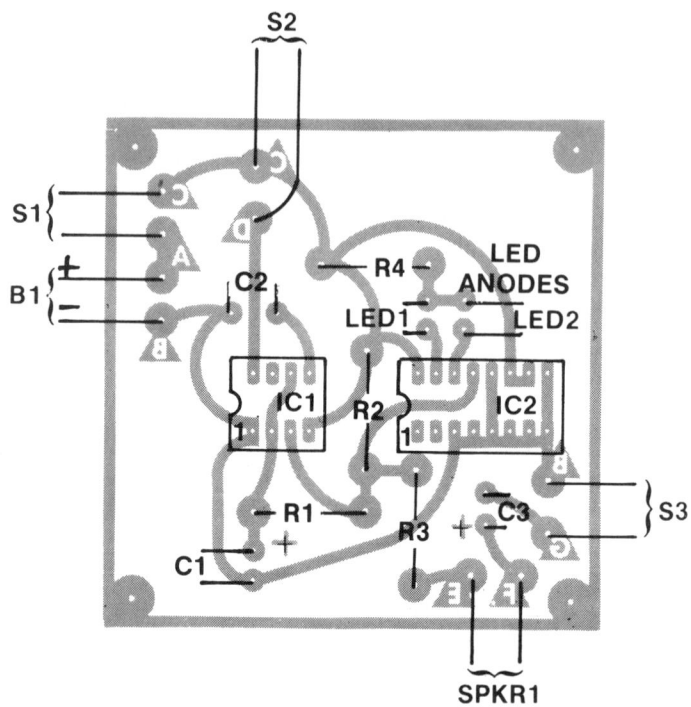

FIGURE 7—6C: ELECTRONIC CRYSTAL BALL COMPO- NENT PLACEMENT GUIDE (TOP VIEW)

Helpful Construction Hints

Insert and solder the component parts in the PC board, carefully noting the polarity of the LEDs, C1, and C2. (See Figure 7—6C.) Sockets are recommended for IC1 and IC2. Use of a low wattage iron is suggested for soldering the closely spaced IC pads to prevent possible lifting and/or bridging of the copper-clad. When installing the ICs in their respective sockets, avoid bending or distorting their fragile leads. To aid in their insertion, a small dot is normally used to designate pin 1. Since IC2 is a CMOS device, it is susceptible to transient voltage damage. S1 must remain open during the insertion or removal of this IC. Complete the project by mounting the LEDs, switches, speaker, printed circuit, etc. in a minibox. (See Figure 7—6D.)

How It Works   (Refer to Figure 7—6A)

Closing S1 applies power to IC2, a J-K flip-flop. When initially powered, one output of IC2 is low (negative) and the other is high (positive). Since LED1 and LED2 are connected to the output pins, they indicate the status of the flip-flop. Depressing S2 supplies power to IC1, a 555 timer, functioning as an astable multivibrator. The timer's square wave output is applied to the clock input of the flip-flop. Each negative going pulse will toggle the logic state of IC2. The values of R1 and C1 determine the multivibrator frequency and thus the rate at which the LEDs alternately flash. Closing S3 provides an aural output of IC1. C3 serves to block the DC output of the timer and R3 limits the current to SPKR1, thus controlling the volume.

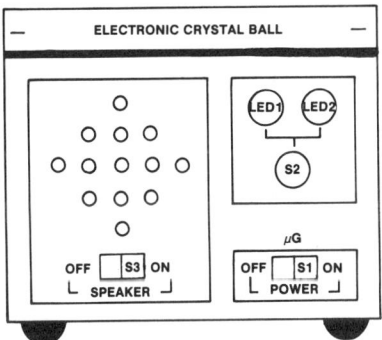

FIGURE 7—6D: ELECTRONIC CRYSTAL BALL (FRONT PANEL VIEW)

Operating Tips

Close S1. Either LED1 or LED2 will light. Momentarily depressing and holding S2 will cause the LEDs to alternately flash back and forth. Releasing S2 will cause one of the LEDs to remain illuminated. Close S3 to provide sound effects when S2 is pressed. After use, be sure to open S1 to minimize battery consumption.

7—7: SOUND MACHINE

What It Does

Here is a project that is guaranteed to be a hit with young and old alike. With the sound machine, you can electronically simulate a variety of interesting and unusual audio effects. Listen to a blaring siren or a tranquil chirping bird. The sounds of old-time steam locomotives and futuristic phaser weapons are yours with the flick of a switch.

Parts to Use

B1 – battery (see text)
C1 – .3 mfd. 25 volt
C2 – .1 mfd. 25 volt
C3 – 470 mmfd. 25 volt
C4 – 10 mfd. 15 volt
IC1 – SN76477 complex
    sound generator
LED1 – jumbo red
Q1 – NPN transistor
    (2N2222A or
    equivalent)
R1 – 2 meg ohm linear
    taper
R2 – 220 K ohm, ¼ watt,
    10%

R3 – 50 K ohm linear taper
R4 – 330 ohm, ½ watt, 10%
R5, R6 – 47 K ohm, ¼ watt,
    10%
R7, R8 – 100 K ohm, ¼ watt,
    10%
R9, R11 – 4.7 K ohm, ½ watt,
    10%
R10 – 10 K ohm linear taper
R12 – 220 ohm, ½ watt,
    10%
S1 – DPDT switch
S2 – SPST switch
SPKR1 – 8 ohm 3-4"
    speaker

Misc. - battery holder, enclosure, hardware, hookup wire, IC socket (28 DIP), knobs, PC board, etc.

Here's the Schematic

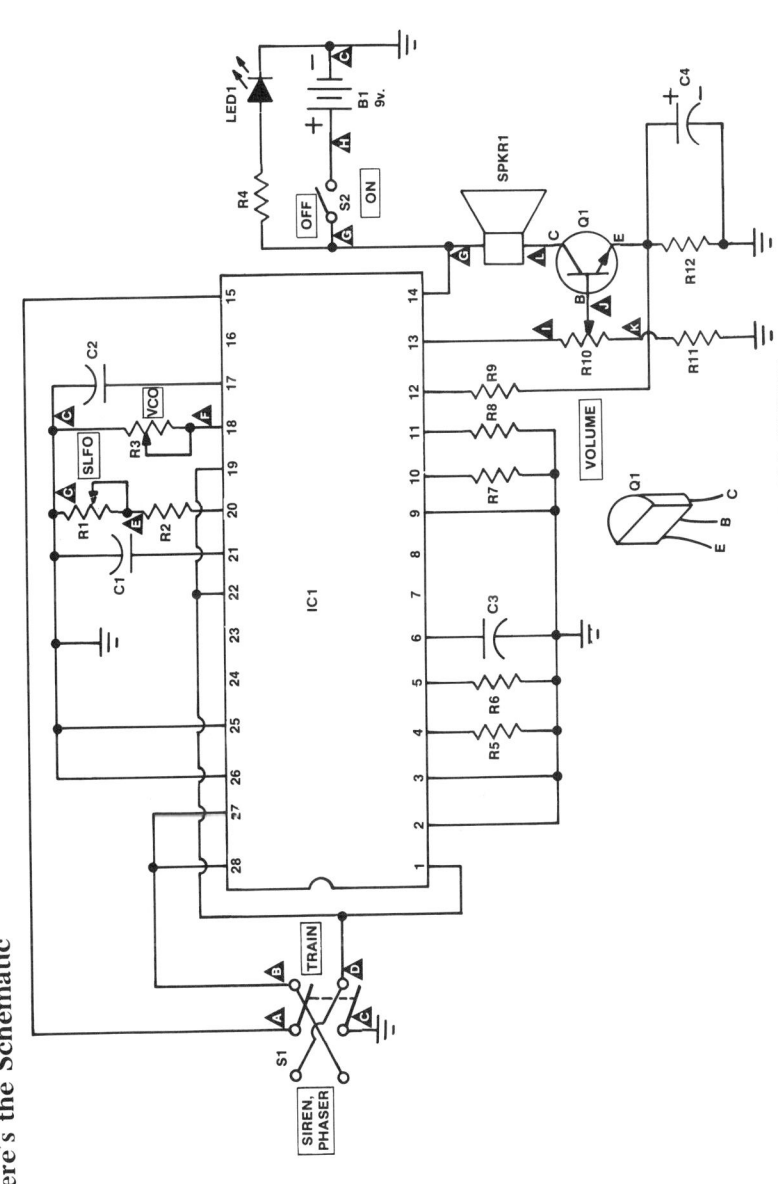

FIGURE 7—7A: SOUND MACHINE

The PC Layout You'll Need

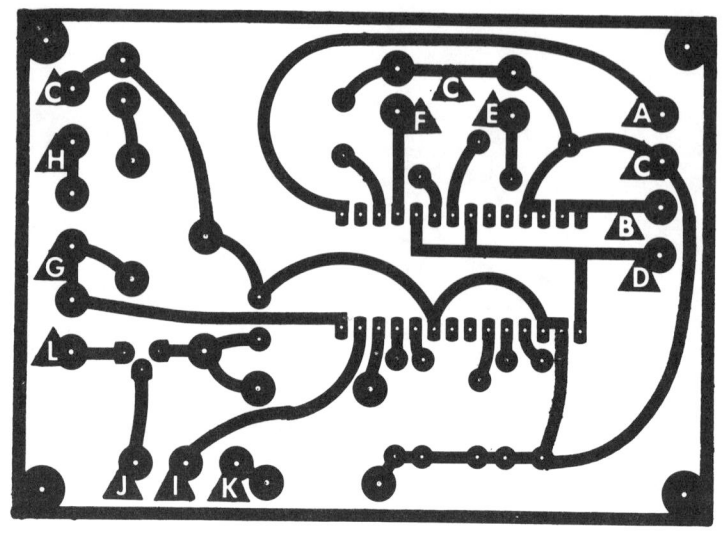

**FIGURE 7—7B: SOUND MACHINE PC LAYOUT
(ACTUAL SIZE)**

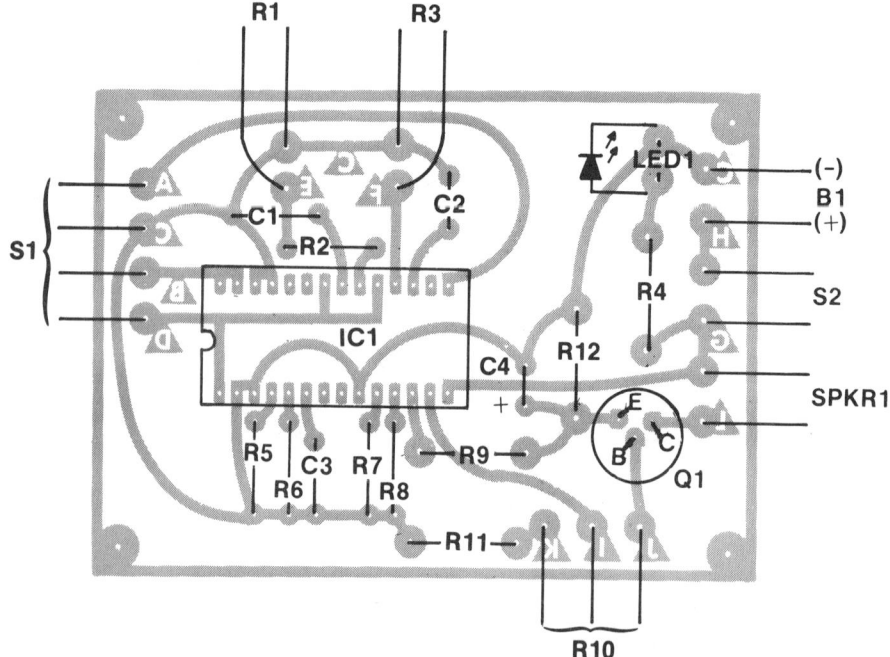

**FIGURE 7—7C: SOUND MACHINE COMPONENT PLACE-
MENT GUIDE (TOP VIEW)**

Helpful Construction Hints (Refer to Figure 7—7C)

Q1 may be any general purpose NPN audio transistor rated at least 500 mw. Carefully observe the lead placement of Q1, particularly if a substitute transistor is used. Avoid possible heat damage to the sound generator with the use of an IC socket. To prevent lifting and/or bridging of the small circuit clad, use of a low wattage soldering iron is suggested. When inserting the IC in its socket, avoid bending or distorting the fragile leads. To aid in installation, a small notch or depression is normally used to designate pins 1 and 28. Because of the high current requirements of the audio amplifier, six series connected penlight cells are preferable to a single nine volt battery.

How It Works (Refer to Figure 7—7A)

The sound machine utilizes a SLFO (super low frequency oscillator), VCO (voltage controlled oscillator), and noise generator incorporated in the sound generator IC. Frequency of the SLFO and VCO are determined by the values of C1, R1, R2 and C2, R3 respectively. R5, R6, and R7 control the operation of the white noise generator/filter. Abling of the various functions is accomplished by logic levels selected by S1. High (positive) levels are derived from the IC's internal 5 volt regulator circuit (pin 15). Low logic levels are obtained from the batteries' negative ground. R11 determines the IC's output amplitude by establishing the operating current of the self-contained operational amplifier. The volume control (R10)

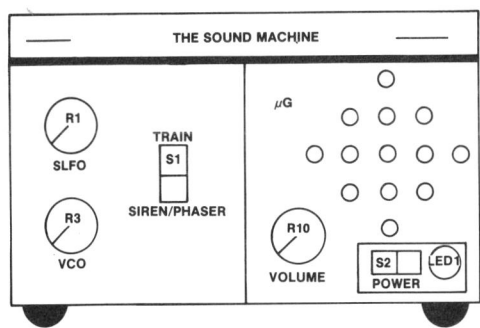

FIGURE 7—7D: SOUND MACHINE (FRONT PANEL VIEW)

regulates the amount of signal voltage from the op-amp applied to the base of the output transistor. R12 and C4 respectively provide emitter biasing and signal bypass.

Operating Tips (See Figure 7—7D)

Closing S2 applies circuit power as indicated by the illumination of LED1. R1 and R3 respectively determine the rate and frequency of the sounds produced. Adjust R10 to achieve a comfortable listening volume. Be sure to open S2 when not using the sound machine to minimize battery consumption.

Siren: Position S1 in the "siren/phaser" mode. Adjust R1 for maximum resistance and R3 to a mid-range setting.

Phaser: Position R3 for mid-range and R1 for minimum resistance.

Chirping bird: Adjust R3 for minimum resistance. Decrease the resistance of R1 to increase the chirping rate—increase R1 to decrease the rate.

Steam train: Toggle S1 to the "train" position. Adjust R1 for the desired locomotive speed.

Index

233